Quality Assurance

A Framework To Build On

Quality Assurance

A Framework To Build On

Second Edition

Terry Hughes MSc, FRICS

Trefor Williams BSc, PGCE, MPhil, MCIOB, MIQA

Blackwell Science

© Terry Hughes and Trefor Williams 1991, 1995

Blackwell Science Ltd
Editorial Offices:
Osney Mead, Oxford OX2 0EL
25 John Street, London WC1N 2BL
23 Ainslie Place, Edinburgh EH3 6AJ
238 Main Street, Cambridge
  Massachusetts 02142, USA
54 University Street, Carlton,
  Victoria 3053, Australia

Other Editorial Offices:
Arnette Blackwell SA
  1, rue de Lille, 75007 Paris
  France

Blackwell Wissenschafts-Verlag GmbH
  Kurfürstendamm 57
  10707 Berlin, Germany

  Feldgasse 13, A-1238 Wein
  Austria

First published 1991
Second Edition 1995

Printed and bound in Great Britain
by Hartnolls Ltd., Bodmin, Cornwall

DISTRIBUTORS

Marston Book Services Ltd
PO Box 87
Oxford OX2 0DT
(Orders: Tel: 01865 791155
            Fax: 01865 791927
            Telex: 837515)

USA
  Blackwell Science, Inc.
  238 Main Street
  Cambridge, MA 02142
  (Orders: Tel: 800 215-1000
                   617 876-7000
              Fax: 617 492-5263)

Canada
  Oxford University Press
  70 Wynford Drive
  Don Mills
  Ontario M3C 1J9
  (Orders: Tel: 416 441-2941)

Australia
  Blackwell Science Pty Ltd
  54 University Street
  Carlton, Victoria 3053
  (Orders: Tel: 03 347-0300
              Fax: 03 349-3016)

A catalogue record for this title is available
from the British Library

  ISBN 0-632-03904-3

Library of Congress Cataloging-in-Publication Data
is available

3

IMPLEMENTATION   27

4

THE SYSTEM DEVELOPED - ITS STRUCTURE AND PHILOSOPHY   41

5

THE SYSTEM DEVELOPED - TENDER STAGE   53

1.0   Background

Within the construction industry it is recognised that an increasing amount of time is being spent in the preparation or examination of contractual claims. It has also become apparent that for a variety of reasons there is a diminishing return from such claims. A key to this is the increased sophistication of clients in applying the requirements of the Standard Forms of Building Contract. These contracts clearly place the burden of proof on the contractor. The task of providing this proof, often in the form of documentary evidence, is frequently complicated by a lack of pertinent records. A failure to promptly provide the required documents disturbs the relationship between contractor and client or more particularly the client's agent. Therefore to make things worse economic loss is often accompanied by loss of goodwill. In addition when tracing the history of some of these disputes it becomes clear that the problems which form the basis of these claims, although frequently materialising at site level, regularly have had their root cause earlier in the management and design process.

Examination of project records shows that questions have not been asked when they should have been, problems are not seen until after they have occurred. It is typically not the case that the individuals involved lack expertise or initiative. It is more often the case that they are pressed for time by other concerns resulting in them giving these issues scant attention. A further difficulty arises where an individual's lack of experience limits their personal horizon and dims their perspective of the problems ahead. Crises caused by poor quality have often overtaken contracts that appeared to be proceeding success-fully, with the resulting delay and cost, undoing the supposed benefits that had been accrued.

It is clear that as the cost of getting it wrong increases, regardless of which party is at fault, the cost of getting it right becomes the sensible economic objective.

1.1   The three factors that management must consider

One of the major problems encountered is an over emphasis by management on time and cost. These are important factors but are only two parts of what is in effect a three-sided triangle of time, cost and quality. These factors must be in balance as any neglect of one will have a corresponding detrimental effect upon the other two. This follows the mathematical rule that an alteration to any angle in a triangle will amend one or both of the other angles. Historically there has been an attempt to manage the construction process by controlling time and cost in progress whilst quality tends to be managed in retrospect. This is understandable since most management control systems examined by the authors highlight time and cost and leave the responsibility for quality to others.

It is our contention that quality in the true sense of the word must be given equal priority with time and cost within contractors' management processes. This is a decision that management must make under the new ISO 9000 model which requires them to give an equal commitment to quality.

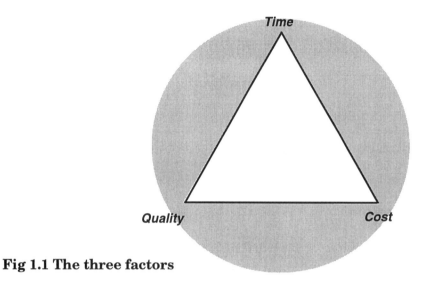

Fig 1.1 The three factors

From this viewpoint of resolving the imbalance between concerns for time, cost and quality an interest in quality assurance developed. At first there was no clear indication of how quality systems could overcome the problems. On first examination quality assurance produces scepticism. Why should the introduction of even more forms and procedures, to an industry already overburdened with them, improve the situation. However continuing interest and research led to the opportunity to assist in the development of a Quality Assurance management system for a contractor's organisation.

The result of this work was an understanding and appreciation of how the contents and process of development necessary for a QA system can in large measure be:

- an answer to many of the problems seen
- not at all the generator of paperwork imagined.

As many managers are aware, frequently by the time site management has become involved in the project the problems are already there, at embryo stage, hidden by a mass of paperwork. A means must be produced to find these problems at this embryonic and less costly stage. QA requires you to define in writing precisely what you do. This very task produces a means of identifying where these problems are likely to occur, highlighting the important from the mundane, the unusual from the commonplace.

Put simply:

QA system means	Results in
* Efficient use of time	* Time available to site manager
* Addressing items in good time	* Time saved in crisis management
* Providing solutions drawn from past experience of others	* Source of 'experience' for site manager
* Placing responsibility where it is intended to be	* Source of records if required
* Generating pertinent records	* A means of measuring performance
* Providing solutions by reviewing corrective and preventative actions taken	* Reduction in replicated errors

1.2   Quality assurance

1.2.1 Why

Commercial reasons

Contracting has always involved the acceptance of risk. The secret of success has been to keep the risk within confined and controllable limits.

The risks currently associated with the discovery of defective work are increasing in both the long and short term.

Long term Latent Damage Act 1986 - 15 years liability with the period of liability running from date of discovery.

Short term The Intermediate Form of Contract introduced with clause 3.13.1. - a new power for architects following the discovery of defects to establish *at no cost to the employer* that no similar failure exists.

This has now been extended to JCT 80 with the introduction of clause 8.4 allowing the architect this same power provided his instructions are *reasonable*. The extent of risk is increasing both in terms of liability, occurrence and cost.

Liability The contractor's share of risk is being expanded from a number of sources:
- legal decisions - duty of care. A tendency toward consumer protection is likely to lead to an expansion of legal liability seen now in the Construction Design and Management (CDM) Regulations

- contract amendments - JCT 80 clause 1.4 states that, not withstanding the architect's obligations, the contractor is *fully* responsible. Pressures on architects via their professional indemnity insurance will continue this trend
- client/client's agents' resistance to contractual claims has meant diminishing returns from this process.

Occurrence The frequency of discovery of defective construction has been highlighted in a number of recent reports.

In the shorter term contractors can expect to find increasing pressure for compliance with better drafted specifications and Government support is likely to expand this to many areas of public expenditure.

Cost It is recognised that the cost of rectifying defects increases in relation to progress. The later the problem is found the more expensive it is to put right. Fig 1.2 could show for example the cost profile of remedying a foundation concrete problem.

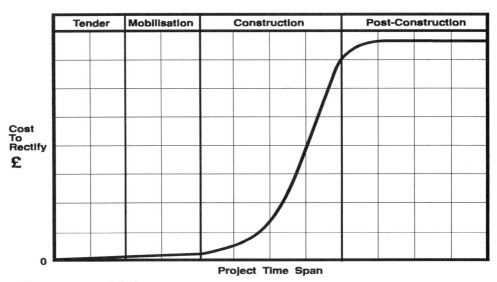

Fig 1.2 Discovery of defect

Marketing reasons

Improved image You can sell on perceived quality, as well as price, Marks & Spencer and Jaguar Cars being prime examples.

Bandwagon Clients are increasingly asking for a company's quality credentials which often equates to inclusion or exclusion from tender lists.

Type of market Ability to tender for work in a more selective market.

Organisational reasons

There are frequently shortages both in terms of quantity and quality of skilled labour and supervision.

Contractors not only have difficulty in obtaining experienced labour but also in obtaining acceptable standards of work from the labour available at the prices payable.

Labour today is generally employed on a labour only basis and is largely motivated by income. A problem associated with this is that they recognise that their income can be increased in four principle ways:

- working longer hours
- pushing up rates per unit of output
- cutting corners to increase output
- working for a well organised main contractor.

Companies should aim to ensure that the final choice of working for a well organised contractor is available. For the above reasons the need for experienced competent supervision has never been greater. Supervision skills are currently, and for the foreseeable future going to be, in short supply. When this is combined with the greater administrative burden that legal obligations and tighter schedules impose the problems of control are exacerbated. A further difficulty apparent within contractors' supervisory structures are the disparate skills required for planning, control and motivation, strengths in the practical skills often going hand in hand with poor administrative skills and vice versa. It should be noted that the new ISO 9000 model requires 'executive management' to ensure the identification and provision of the correct resources, including trained personnel, required for the execution of the work.

1.2.2 How

Options to improve quality:

- improve supervision and inspection
- build-in quality using a systems approach.

Improve supervision and inspection

Traditionally quality control has involved inspection and testing as the work proceeds. Although increased supervision and inspection can improve the performance and quality on site there are several disadvantages of this option.

1. Increased cost of supervisory staff.
2. The nature of building sites makes it almost impossible to inspect every stage of the work.
3. This option is not always able to predict difficulties but has to provide solutions as and when they arise (management by crisis).

4. Testing is retrospective and thus costly to remedy faults.
5. For some 'special' processes it may not be possible to verify quality simply by increasing inspection.

Furthermore evidence in manufacturing industry has shown that even extremely sophisticated inspection measures in a controlled environment have only a limited and temporary success. Increased inspection and supervision is really a 'negative culture approach'. It is to be equated with fire fighting and is often no more than a matter of recording what is wrong and accepting its cost.

Systems approach

An alternative to increased supervision is to adopt quality systems. These systems provide a way of working that either prevents problems arising or identifies and deals with them effectively and cheaply if they arise. The emphasis is on fire prevention rather than fire fighting.

Addressing these problems early enables management to be proactive. Systems identify when tasks are being carried out correctly or whether they have been carried out at all. They allow a manager to manage, but more importantly are a means of directing employees towards what needs to be done.

1.2.3 What

Quality was defined in BS 4778 : Part 1 : 1987 as *The totality of features and characteristics of a product or service that bear on its ability to satisfy stated and implied needs.*

In the construction industry the technique, known as 'Quality Assurance', had previously only been applied to high value projects where public safety and integrity of construction was of paramount importance, e.g. nuclear installations, offshore structures, etc. Outside these sectors of the industry there had been little or no commercial incentive for professionals and contractors alike involved with the industry to introduce QA. However, since the publication of a Government White Paper in 1982 entitled 'Standards Quality and International Competitiveness', followed by a National Quality Campaign, greater impetus was given to the wider use of the concept and its development into more commercial areas.

Beneath the jargon-ridden mystique the principles of QA are really quite simple. These are to devise methods of doing work systematically so as to avoid errors and costly mistakes. A QA system has been commonly described as:

- saying what you do
- doing what you say
- recording that you have done it.

Saying what you do

This is a simple statement which belies the very complex task necessary to achieve it. The statement 'say what you do' can be looked at from different perspectives each of which would consider that statement in differing depths e.g. the managing director's overview is not always the same as the employee's perspective of the task. A good QA system should therefore view this statement from the perspective of those who have to undertake the tasks described.

Doing what you say

This is an action following the written statement of your intentions in the preceding item. It should be appreciated that the greater the detail used to describe what you do, the easier it should be for the individual to follow it. However this should not try to replicate by written procedure every nuance of skill or craft training which may be involved. In tandem with this it is important that the detail is a faithful description of how various tasks are executed. If not it will soon become apparent that the system is merely what you wish would happen and not what does happen, thereby devaluing its usefulness. There is also a danger that too much minute detail of methodology could lead to auditors raising non conformances when an individual's approach varies slightly from the norm set by the system.

Recording that you have done it

Having both defined and performed the tasks it is necessary to produce evidence that this has in fact occurred. Such records are useful to the individuals that have carried out the task, their superiors who have the responsibility for ensuring the task is undertaken and to third parties who require the provision of such evidence.

1.2.4 Advantages and disadvantages of systems approach

Advantages derived from available research data are:

1. Improved communications and efficiency.
2. Checking of work and avoidance of unnecessary and costly errors, failures and expensive remedial works.
3. Documented proof that work has been executed in compliance with the document and specifications.
4. Easier implementation of client changes.
5. Precise clarification and quantification of the effects of such changes.
6. Easier identification and quantification of delays and claims.
7. Completion on time.
8. Reduced maintenance period remedial works.
9. Provision of as-built records.
10. Possible reduction of insurance premiums.
11. Improved competitiveness and marketability of services.
12. Documented methods that clarify and thereby reduce training time.

Disadvantages

There are three perceived problems which have been identified:

Need to retain records Can be costly and take up space. The precise quantity of information will vary due to the requirements of individual systems and the nature of the task. The system is required to define the extent, nature and duration of record retention.

Additional staff requirements Experience seems to indicate that additional staff are required during establishment and implementation. If there are any additional staff costs they will probably be far outweighed by the eventual long term benefits previously outlined.

Start-up costs and certification charges Precise figures are obviously difficult to obtain but research seems to suggest that full establishment of systems with third party certification would represent about 0.3% of one year's turnover over a period of two years i.e. 0.15% each year.

1.3   What a QA system is trying to do

It is the production of a checkable written guide for the use of individuals. These include employees, managers and third parties.

It produces a template for the individual employee to use in performing their particular tasks, that the manager can use in ensuring the proper performance of those tasks and that the client or third party can use to confirm that their requirements have been met.

Managers benefit as individuals performing their own tasks and are assisted in their leadership function. The definition of subordinates' and colleagues' roles and responsibilities together with lines of communication and evidence of performance provides a clearer means of control.

Third parties, even more than managers, appreciate they cannot check that all tasks have been executed in accordance with their requirements. The system gives them this assurance and overcomes the uncheckable nature of some tasks. An extreme example of how this confidence operates is the American Corps of Engineers who rely for quality solely on the assurance provided by their QA certified contractors. There is however a sting in the tail of this apparent blind faith in that should any contractor be found to have not complied with the requirements of their QA system they may immediately be precluded from any future work.

Individuals that we have interviewed who have used such a system have appreciated its benefits in that the system as a whole describes both the organisation, including their roles, and the tasks they are required to perform. A common statement from them was that the system enabled them on joining the organisation or project to know what their roles were and what was expected of them. They can quickly fit in and feel involved.

1.3.1 Implementation

For any QA system to be meaningful it must include assessment and certification. There are three main grades of QA system each related to the choice of certifying party. Thus you can have first, second or third party certified systems each with varying degrees of acredibility and thus acceptability, although to have any credibility at all systems must comply with ISO 9000. This issue is explained in more detail in Section 3.0.

In addition the level of certification sought can vary:

- to include or exclude design i.e. ISO 9001 includes design, ISO 9002 excludes design
- to set a minimum value e.g. the system will only apply on projects exceeding £1M
- to be applicable to specific project type only e.g. housing.

Certifiers stipulate that a third party system must be applicable to a significant part of the company's operations.

The grade and level of certification chosen may be subsequently extended by expanding the coverage of the original system developed ie the scope of operation that the registration covers.

The need for a QA system can develop either from a second party's requirements or from the recognition by management of inherent problems within the company's organisation. Whichever is the source it is necessary if the organisation is to develop a factual rather than a fictional system that a belief and commitment to producing quality exists. A QA system is not an 'optional extra' only to be used when required but is a philosophy which must be totally integrated with the company's existing management systems. Unless the top management believe that QA will provide a benefit to their organisation then it is unlikely that the system subsequently developed will actually achieve all that it is intended to. A commitment to QA will invariably mean changing the way things are done at present and require that discipline is imposed on the whole of an organisation from 'executive management' down.

Experience and research have indicated that in producing a QA system organisations tend to encounter two major problems:

1. Not appreciating the volume of work required to prepare a system e.g. they will appoint an individual from within the organisation to prepare the QA system, frequently in addition to their other duties. This leads to one of two results:
   - the task becomes elongated leading to a loss of impetus or
   - the individual will address the problems in less depth than is actually required, producing an ineffective system.

2. Balancing the need for both a knowledge of QA and an understanding of the functioning of the organisation. This is best illustrated by the continuum shown below:

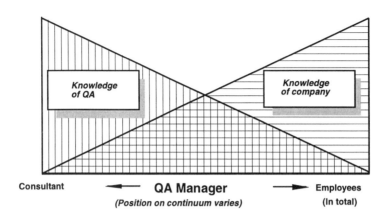

Fig 1.3 Knowledge continuum

The use of QA consultants in most cases tends to span a relatively short proportion of the total period required for the preparation of a system. Also it tends to occur at or near the start of that process. Whilst the QA consultant will provide you with the basic philosophy of a QA system they often lack knowledge of the tasks that such a system must address. This knowledge rests very much with the totality of individuals within the organisation. Managers can give you the overall picture of the tasks that are undertaken. Only the individuals responsible can describe how those tasks are executed in sufficient detail for your analysis. The problem is reconciling knowledge of the company and quality assurance at the right time i.e. the QA consultant's advice is frequently received before a full appreciation of the scale of the problems is available. The appointment of a QA Manager provides an interface between these sources although his own knowledge of either aspect will vary. A more detailed discussion of the role of the QA Manager can be found in Chapter 10.

Our book seeks to provide some remedies to these problems. It will firstly provide a skeleton to substantially reduce the hundreds, possibly thousands, of manhours necessary in establishing relatively common processes and procedures. Secondly it enables the company to look ahead at an early stage when the QA consultant is still closely involved. This knowledge will enable a more effective use to be made of a relatively expensive asset.

The next chapter is a synopsis of BS EN ISO 9002 : 1994 and relates the terminology used therein to the construction process.

2.0   General

As noted in the previous chapter, a company's QA system should comply with ISO 9000. This International Standard is produced in four parts:

ISO 9001   Specification for design/development, production, installation and servicing.

ISO 9002   Specification for production and installation.

ISO 9003   Specification for final inspection and test.

ISO 9004   Parts 1 - 4 Quality management and quality system elements.

ISO 9002 : 1994, excludes design and is therefore the appropriate choice for controlling a construction company's traditional activities. A contractor's design and build projects can also be covered under this International Standard provided the design is bought-in. Otherwise compliance with ISO 9001 : 1994 would be necessary which would add the necessity to address the requirements of Clause 4.4 Design Control.

2.1   Philosophy of ISO 9000

The philosophy is that in order for a company to be successful it must offer products or services that:

1.   Meet a well defined need, use or purpose.
2.   Satisfy customers' needs and expectations.
3.   Comply with applicable standards, codes and specifications.
4.   Comply with statutory (and other) requirements of society.
5.   Are made available at competitive prices.
6.   Are provided at a cost which will yield a profit.

The means by which this is achieved is through a system of quality management which aims to 'get it right first time'. Client assurance requires demonstration of an organisation's capability to control the processes that determine the acceptability of the product supplied. The requirements of the Standard are aimed primarily at preventing and detecting any nonconformity during production and installation. This includes the implementation of a means of preventing its recurrence. These points are emphasised in the Standard as a means of achieving customer satisfaction.

It is applicable in contractual situations when:

- the specified requirements for the product are stated in terms of an established design or specification. Two definitions of specified requirements are given within ISO 9001 and ISO 9004-1.

- confidence in product conformance can be attained by adequate demonstration of a certain supplier's capabilities in production and installation.

Whilst the above statement is of course true you will find in practice that it is difficult to merely use these principles on a selective basis only. The process of development of a workable QA system explores many facets of an organisation's existing procedures and structure. Existing company procedures will almost certainly be amended during the development of a QA system and the likelihood of the company devising two new procedures, one to operate with QA and the other without, is remote.

2.2   Synopsis of ISO 9002 : 1994 clauses

Figure 2.1 is a diagramatic representation of the requirements of ISO 9002 : 1994 Quality Systems. The shadowed boxes represent the majority of the clause numbers, with the balance being shown as links between the various elements e.g. Traceability Clause 4.7. The shadowed boxes are linked to represent the interrelationship of each of the clauses within a quality system. The quality system itself, Clause 4.2, is shown double boxed as it represents the repository of the information shown on the remainder of the diagram. In addition to the shadowed boxes and their links, other boxes are included to represent other elements found within the construction process but not specifically referred to in the British Standard. Although the system operates throughout the company the diagram also indicates that certain elements are primarily project specific.

The synopsis hereafter gives a brief interpretation of each of the clauses within the International Standard and explains how the diagram links these various elements together.

Each of the clauses noted on diagram Fig 2.1 is examined as follows:

- each page starts with a brief interpretation of the essential features of the clause

- below this is an extract from the diagram shown in Fig 2.1. together with a description of that element and its relationship to the other clauses

- finally there may be some additional notes of explanation.

For further information refer to ISO 9001 : 1994 and ISO 9004 : 1994.

Authors' note
ISO 9000 uses the terms Supplier, Customer and Subcontractor. However, their meaning varies from that generally used in the industry. For the ISO definition of these terms refer to page 25.

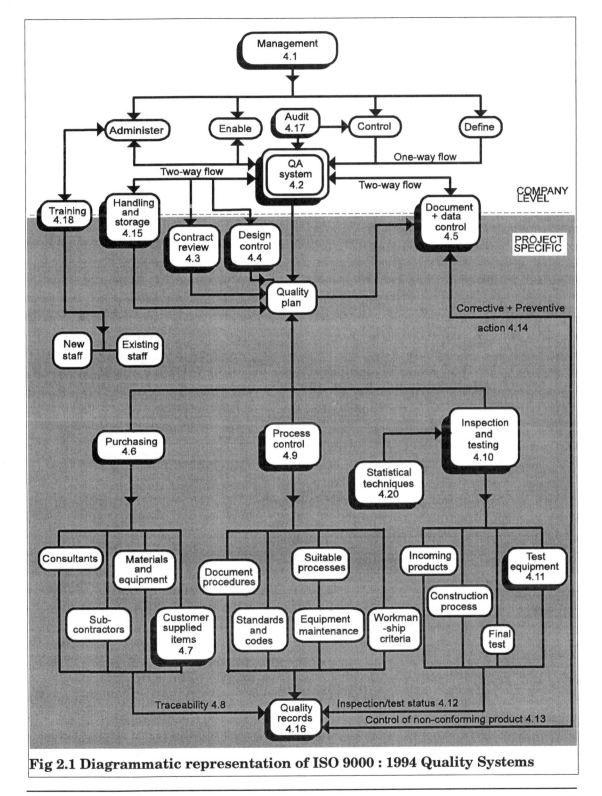

Fig 2.1 Diagrammatic representation of ISO 9000 : 1994 Quality Systems

2.2.1 Management (Clause 4.1)

Executive management of the company (note - the Standard refers to the company as the Supplier) must:

Define

State its policy, organisational goals and objectives which include its commitment to quality and its intentions relevant to satisfying the expectations and needs of the customer. This policy must be understood, implemented and maintained at all levels.

The quality system must be implemented from executive management down under the guidance of the senior executive manager and a quality representative who must be appointed from within executive management.

Enable

Delegate in writing defining the responsibility, authority and role of all personnel who manage, perform and verify work affecting quality.

Delegate in writing, authority to stop or reject work and the power to take action to prevent repetition.

Administer

Identify and provide adequate resources for managing, performing work and verification activities (including internal quality auditing). Ensure that the personnel noted above have the necessary experience, qualifications or have received appropriate training commensurate with the responsibility and role undertaken.

Control

Appoint or nominate a manager (who must be an employee of the company) with defined authority and responsibility for ensuring that the requirements of the Quality System are implemented and maintained, and its performance reported at Management Review.

Regularly review the system at specified intervals to ensure its continuing effectiveness.

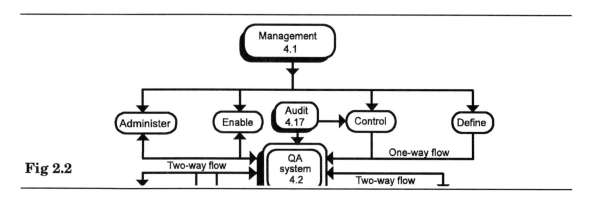

Fig 2.2

Fig 2.2 shows that management is directed via the Quality System, which management establishes and maintains via regular feedback. They are specifically required to consider the quality policy and objectives at these times. Control is exercised via Audit Procedures, see 2.2.11 later, which is a two-way process of continual improvement for the Quality System.

2.2.2 QA system (Clause 4.2)

- The system must be defined in a document which must include a Quality Manual.

- The document must be available to all personnel who have defined roles and responsibilities within it.

- The document must define the organisational structure, responsibilities, processes and resources applicable to ensuring that the project conforms to the specified requirements.

- Quality Planning is a requirement of the Standard and should be consistent with the Contractor's QA system. The Quality Plans prepared should relate to the specific requirements of the contract.

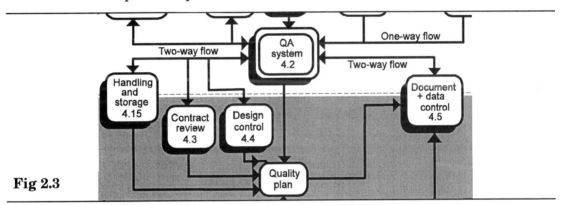

Fig 2.3

Fig 2.3 shows the QA system at its heart with its constituent elements feeding both to and from it. In a construction organisation the system exists at two levels. At company level are those processes and procedures which management require to be common to the whole organisation. At project level are the specific processes and procedures to operate in the unique circumstances of the project. The system must firstly be well planned and managed and secondly must produce documentary evidence of its existence and effective use. The Quality Plan is a vital element of the system but need only be as detailed as the complexities of each particular project demands. The Quality Manual is in effect an outline of an organisation's policy and the way it conducts its business. Quality Procedures provide the detailed methodology of how business activities are performed.

Authors' note

In practice the document also exists at two levels:

1. *The Quality System which will include the Quality Manual and Quality Procedures which will be generally applicable to all relevant projects and operates at both company and project level.*
2. *The Quality Plan which BS 4778 defines as a document setting out the <u>specific</u> quality practices, resources and sequence of activities relevant to a particular product, services, contract or project and operates <u>specific</u> to that project in conjunction with the Quality System.*

2.2.3 Contract review (Clause 4.3)

A procedure must be included at both Tender and Pre-Commencement Stages to review the client's requirements and:

1. Ensure that they are adequately defined, specified and documented with procedures for confirming verbal instructions.
2. Define and resolve any ambiguities or contradictions.
3. Identify and process amendments.
4. Ensure that resources are available.
5. Document the results of the above (Records of Contract Review).

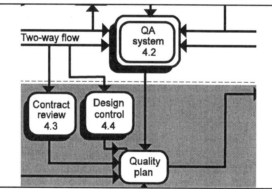

Fig 2.4

Fig 2.4 shows that the procedures to be adopted within the review noted above are defined within the system. Also that the documented results specific to the project are a necessary element to assist in the compilation of the Quality Plan. The process is intended to provide a clear recorded definition of each party's intentions or requirements at an early stage of the project.

Authors' note

This procedure is essential to the proper preparation of an effective Quality Plan. It also assists all parties in the identification of design / construction related problems at an early stage when experience suggests an amicable solution can often more easily be agreed.

2.2.4 Design Control (Clause 4.4)

This clause is required in all Quality Manuals but for organisations working to ISO 9002 this subclause simply needs a note to state that it is not applicable. The purpose of this clause is to standardise numbering for all ISO 9000 systems.

2.2.5 Document and Data Control (Clause 4.5)

Procedures must be included to ensure that all drawings, specifications and other quality related documents and data, including revisions, are properly controlled.

Drawing and other registers must be kept up to date and record to whom copies have been distributed.

Procedures should be included to prevent the use of out of date information. Copies of superceded documents should be retained but clearly marked to prevent accidental use.

The procedures should extend to the control of information feedback in the form of Quality Records.

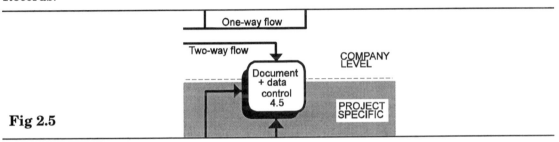

Fig 2.5

Fig 2.1 illustrates the position of document control as we believe it should be used. The QA system establishes the procedures to be used and the Document and Data Control system feeds those records relevant to the maintenance of the system back to the system to be available for inspection and audit. These records also assist management in its task of pursuing the continual improvement of the QA system noted in Clause 4.1. Documents and data can be in a variety of media forms including hard or floppy disks, tapes, microfiche, photos, paper copies and other electronic means.

Authors' note

The control of documentation and data is central to the effectiveness of a Quality System and will provide a major source of information for audits of the system. This control is ensured by:
1. *Responsible personnel being identified.*
2. *The distribution being defined and recorded.*
3. *The documents or data themselves being adequately referenced so that they can be traced and identified.*
4. *A master list being readily available.*
5. *The removal of invalid as well as obsolete documents.*

2.2.6 Purchasing (Clause 4.6)

Procedures, graded in stringency to the importance of the purchased item, must ensure that bought-in resources comply with the specified requirements and that those providing them are competent. Such compliance should be judged by:

1. Evaluation of their formal Quality System.
2. In-house records of previous achievement.
3. Formal internal evaluation procedure for examination of physical or documentary evidence of past performance.
4. Provision for higher level of main contractor supervision and support of their quality standards.

Procedures should also be included for the continual re-evaluation of bought-in resources based upon records of performance.

Procedures must be included to ensure that customer (client) supplied services or products, including information, also comply with the appropriate quality standard. (Clause 4.7)

Also include procedures that enable the standards and sources of products incorporated within the works to be identified and traced. (Clause 4.8)

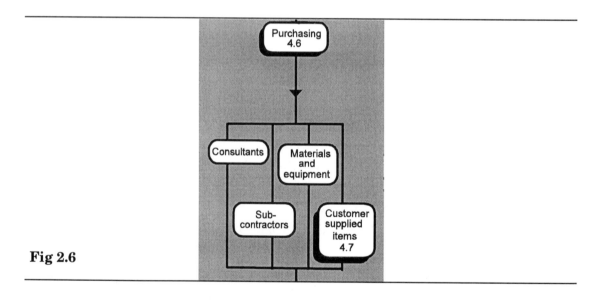

Fig 2.6

Fig 2.6 illustrates the nature of the resources to be purchased including those supplied by the client. The traceability element shown on Fig 2.1 is fed back to the system via the quality records and document control.

Authors' note

Emphasis should be placed on the following:

1. Review of approval of purchase documents to ensure the customer's (client's) require-ments are satisfied.
2. Subcontractors, which includes suppliers, are evaluated within a <u>scope of supply</u> ie suitable for product or service A but may not be for product or service B.
3. Maintaining a record system for subcontractors (and suppliers) which shall be reviewed on a regular and defined basis.
4. Judgement must be used in applying these procedures and should be commensurate with the specified requirements or access for later inspection i.e. buried work may entail more stringent procedures than work which remains visible and open for inspection.

2.2.7 Process Control (Clause 4.9)

Procedures must be included for:

- planning
- subsequent control
- documentation.

The procedures may be graded in rigour dependent upon judgement of the item's impact on quality.

Written method statements or instructions must be included where product quality would be affected if not used.

Intermediate stages must be defined, supervised, controlled and documented where work cannot be properly assessed in its final state.

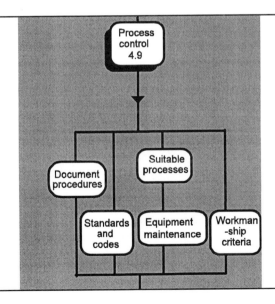

Fig 2.7

Fig 2.1 shows Purchasing, Process Control and Inspection and Testing all feeding into the project specific Quality Plan. It is therefore within the Quality Plan that the documented project specific details of control of these elements will be produced. Fig 2.7 shows those elements of process control normally required to be catered for. Written method statements or instructions should be recorded within the Quality Records which should include details of equipment maintenance.

Authors' note

1. *These items, where project specific, are identified within the Quality Plan. However where instructions or written method statements are required these could frequently take the form of standard checklists or procedures.*
2. *Maintenance of equipment should also include essential vehicles and machinery used in or support of the process including delivery vehicles.*

2.2.8 Inspection and testing (Clause 4.10)

Procedures must be included for:

- verification of compliance with specification. These must be produced at each of the following stages:
  1. upon receipt
  2. during construction
  3. at completion
  Note - records must be kept as evidence of compliance. The procedures must identify the person with authority to accept conformance.

- the routine checking and documentation of measuring and test equipment (Clause 4.11)

- the prevention of defective materials or workmanship being finally incorporated into the works through a system of checks and records (Clause 4.12) and the control of non-conforming products (Clause 4.13)

- the implementation by management of corrective and preventive action in the event of non-conformance, from whatever source, and evidence of analysis of the reasons for the non-conformance and its correction (Clause 4.14)

- establishing the need and use of statistical techniques in establishing conformance (Clause 4.20).

Fig 2.8 links Statistical Techniques, Equipment Tests, Inspection and Test status and Non-conformance control within the area of Inspection and Testing. All of these elements would form part of the Quality Plan and the procedures for its preparation are given special consideration within Chapter 8.

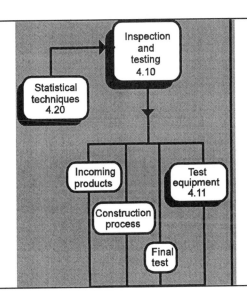

Fig 2.8

Authors' note

The use of statistical techniques within this International Standard when applied to construction would normally be extremely limited e.g. to establish the necessary frequency of concrete cube tests.

2.2.9 Handling, storage, packaging, preservation and delivery (Clause 4.15)

Procedures must be included for:

- receipt of materials

- identification of materials

- inspection of materials

- handling and storage of materials in accordance with the manufacturers' recommendations to ensure protection from damage or deterioration

- preservation (protection) of completed work until handover.

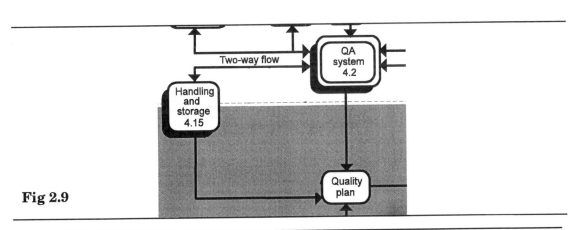

Fig 2.9

Fig 2.9 indicates that a number of the above procedures will be common to all projects but that particular circumstances may well require project specific procedures which would then be identified and written down within the Quality Plan.

Authors' note

Provision should be made within the procedures to prevent the use of materials which have exceeded their shelf life and thus the inspection of materials noted above must include periodic inspection of stored materials.

2.2.10 Quality records (Clause 4.16)

Records must be maintained, safely stored and be accessible for verification that work has been carried out in conformance with the quality system.

The degree of documentation and retention times of records shall be that agreed with the client or defined within the Quality System.

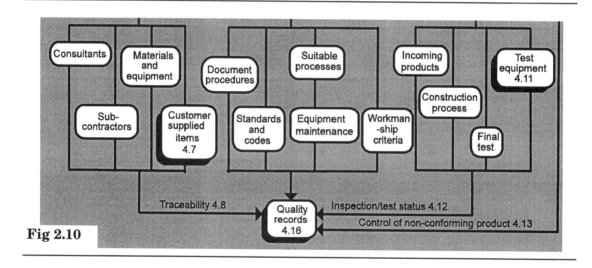

Fig 2.10

Fig 2.10 illustrates the wide variety of sources from which the Quality Records are required to be drawn. Any meaningful system must aim to make full use of these records as they form an important element of the feedback necessary for management's role in the continuous improvement and upgrading of the Quality System.

In addition to the areas shown in Fig 2.10 the Standard requires that the following areas should also be covered within the quality records:

Contract Review
Management Review
Quality Plans
Internal Quality Audits
Training

Authors' note

The records kept should be sufficient in terms of type and extent to demonstrate the required achievement of quality and should include pertinent subcontractors' quality records.

2.2.11 Internal Quality Audits (Clause 4.17)

A planned sequence of internal audits must be defined to ensure the effectiveness of the Quality System.

This is a formal procedure by a delegated properly trained individual independent of the project structure and reporting directly to executive management.

The results of audits must be documented and the records should indicate deficiencies, corrective and preventative action and its timing, the person responsible for such action, and an assessment of its effectiveness.

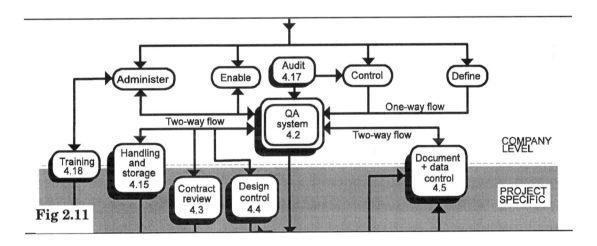

Fig 2.11

Fig 2.11 shows audit as the means of management control via the QA system and in particular the Quality Records fed back via document control to that system. It must be remembered that the audit is concerned with the operation of the system; the responsibility for ensuring that the required quality standards are attained, through use and improvement of the system, rests with management at all levels. The link to executive management control is also clearly indicated in order to facilitate an effective management review.

Another function of audit not clear from the diagram is to check the adequacy of training.

Authors' note

Such audits would be independent of a third party certifier; however records of their occurrence and any remedial action taken would be maintained and would provide evidence of the system's effectiveness.

2.2.12 Training (Clause 4.18)

Procedures must be included to ensure that:

- personnel records are kept and are available to confirm that staff or operatives requiring particular skills have been trained, tested or otherwise checked
- provision is made for additional training where a general or specific need is identified
- provision is made for quality awareness training
- particular attention is given to new personnel.

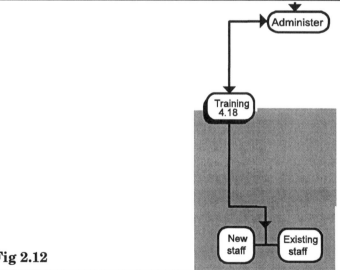

Fig 2.12

Fig 2.12 shows executive management's direct input into the improvement of what is undoubtedly its most important long term resource. The development of this resource concerns management at all levels of the organisation. Procedures for training, although primarily operating company wide, can also be project specific if the needs of a project introduce particular training requirements.

2.2.13 Servicing (Clause 4.19)

This procedure is for after sales service. If this is not appropriate to your company this should be clearly stated in the Quality Manual.

2.2.14 Statistical techniques (Clause 4.20)

The Standard requires that you identify the need for statistical techniques and have procedures for controlling their use.

Authors' note

ISO 8402: 1994 provides the following definitions:

Customer - *recipient of a product provided by the supplier.*

Supplier - *the organisation which provides the product by a process which satisfies the requirements of the quality management system (ie the organisation seeking registration).*

Subcontractor - *an organisation, or individual, providing service(s) or material(s) to the supplier.*

The next chapter describes in general terms how you would implement a quality assurance system within a construction organisation.

This page left intentionally blank

3.0　Generally

Looking at the synopsis in Chapter 2 it can be seen that the ISO 9000 even when interpreted provides only a skeletal outline of the basic features that a complying QA system must provide. The difficulty then arises in applying these broad aims across a wide spectrum of activities and ensuring that the means are provided for their control. Regardless of the solution found, the process of implementation will pass through distinct stages. The three identified are:

- investigation
- development
- testing

although these stages often overlap.

In practice there is also a less definable and earlier fourth stage of instigation. This stage commences with a realisation of the need for the management of quality within an organisation and concludes when a decision to proceed with a quality system is taken. At some point there is a crossover from instigation to investigation and although often the activities can run in parallel, it is preferable that a firm decision to produce a QA system be taken prior to the commencement of the investigation stage.

On conclusion of the instigation stage it is necessary to decide on the type of quality assurance system the company requires, and as previously mentioned there are three distinct options available:

First party certified

The company defines and polices its own quality system and thus has more control over its content and operation. A major disadvantage is that the system generally lacks credibility.

Second party certified

A company's quality system is approved and assessed by your customer who is the other party to the Contract. This can be a problem for companies operating with a large number of clients and also for clients who have to assess each individual contractor's system. Examples of client organisations that have operated such systems are the former Central Electricity Generating Board, British Nuclear Fuels Ltd. and various petro-chemical companies.

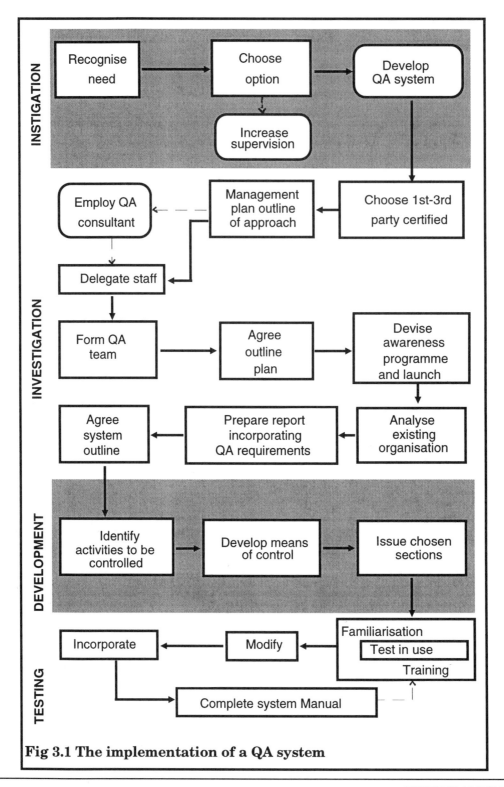

Fig 3.1 The implementation of a QA system

Third party certified

The system is authorised, checked and audited by an independent body allowing the company less control over its content and operation but ensuring discipline and credibility. Accredited Certification Bodies are listed on page 40.

Whichever system is developed it can, if desired, be restricted to operate within a part of the company's activities only, although as noted previously for third party certified systems, that part must be a significant portion of the company's activities.

Timescale

The implementation of any QA system will involve steps similar to those shown in Fig 3.1. A by-product of the system's use is improvement and therefore change, which means that the last four steps contained in the Testing stage must follow a cyclical process.

A minimum period for the implementation of the initial system would probably be nine to twelve months.

3.1 Instigation

As previously discussed this initial stage commences with a perceived need to assure the achievement of quality. Next there is an examination of the options available. As stated in Chapter 1 the choice lies between increasing supervision and the implementation of a Quality System. Once a decision to produce a QA system is taken the process passes from the instigation to the investigation stage. It is at this point that 'executive management' must commit themselves to the process to ensure success.

3.2 Investigation

The first step at this stage is to decide the grade of system to be produced. The choice is a system that satisfies either:

- the company's internal requirements (1st party)
- a particular customer's requirements (2nd party)
- all potential customers' requirements (3rd party).

Having decided this it is necessary to plan in outline an approach suitable to the company's particular circumstances. The next decision is whether to employ a QA Consultant. It is worth noting that some funding may be available through schemes such as TEC/Business Link for part of the cost. Alternatively the choice may be to handle the entire process in-house (note the continuum Fig 1.3).

Having thereby formed the QA team they, together with management, must then agree the plan for the balance of the investigation, development and testing stages of the system.

A quality awareness programme should be planned and implemented. This programme is intended to notify all staff of the company's objectives and should be designed to provide reassurance and foster their continuing support during the three phases of its implementation.

Various methods can be used but it is essential to emphasise top management's total commitment by for example their presence at a launch gathering of personnel at some special venue. The aim must be to give this process a high internal profile from the very start. Following the company's launch the implementation of the next three stages may then proceed.

3.2.1 Analyse existing organisation

In our opinion it is fundamental for the long term success of the system that it be developed as an integral part of the company's existing procedures. As a pre-requisite to this development it is therefore essential to examine the company in great detail in terms of its organisation and procedures.

For this process there are at least three methods which can be used:

- firstly, self analysis by section/department heads and their staff
- secondly, analysis by QA Manager appointed from within the company's management
- thirdly, analysis by external consultants.

Each method has particular advantages and disadvantages (see Fig 1.3). We believe that a combination of all three can maximise the advantages and minimise the disadvantages of each, producing a blend of understanding of QA with a detailed knowledge of the organisation's activities.

This analysis may appear to be an unnecessary exercise as every organisation will already have a detailed knowledge of what it does. However, the company's perception of its organisation and structure will on closer examination normally prove false.

Typically this examination will reveal:

- titles and positions that often do not match the actual roles undertaken
- procedures that in practice are frequently ignored or modified for a variety of reasons
- procedures that are frequently out of date or contradict other stated principles within the same or other documents.

A company's organisation and procedures have often been developed on an ad hoc basis and the introduction of a QA system has the additional benefit of providing the opportunity to rationalise the whole.

The introduction of a quality assurance system enables a company to undertake a critical review of current or previous methods of working, and is of itself often a process of improvement.

In examining this actual structure you must show roles and responsibilities, methods of control, routes of information flow and the means by which these operate.

To gain this information the following means may be used:

- structured interviews using a standard format of questions that will produce or confirm the information required

- questionnaires again using the standard format but questions must be unambiguous

- study of documentary evidence enabling comparison to be made between what has been said and what has happened.

Information via these methods will need to be gathered from all areas of the company including central support departments, project teams, company manuals and a detailed examination of records of current and past projects.

Interviews

These may be appropriate for senior or key personnel and they have the advantages of:

1. Guaranteeing a prompt response to your enquiry.
2. Giving an opportunity to ask secondary or follow up questions if the response is unclear.
3. Assuring and relaxing the person answering who is more likely to open up.
4. Assisting in the development of the questionnaire.

The obvious disadvantage is the time commitment of both interviewer and interviewee.

Questionnaires

They will normally follow interviews (see 4. above) and may be targeted to junior or subsidiary personnel. It is unlikely that a single questionnaire will be appropriate to all employees, so modified versions may have to be produced for specific groups or individuals although the central core of information is likely to remain the same. Questionnaires have the advantages of concurrence of completion and time saving for interviewers. However the disadvantages almost mirror the advantages of interviewing:

1. They need following up to get both a complete and prompt response - it helps if responsibility for ensuring this is given to their superior following that person's interview. A timescale for completion must be given.
2. The questions must be as clear and unambiguous as possible- a pilot exercise is advisable.
3. It helps if they are returned in pre-marked sealed envelopes and some confidentiality is provided.

If the information so provided is found to be of interest or raises further questions this can then be followed through by means of an interview.

Not all responses will be accurate and the information will need to be checked. The spread of responses will enable development of the general pattern of information required.

Study of documentary evidence

This is the final source of information and is most useful to fill in gaps or check the accuracy of other information. It is also likely that information from two sources will conflict; these conflicts are often best resolved by bringing both sources together and discussing the problems which may only be a matter of perception or interpretation.

It is however important during this stage to take care to avoid alienating any individuals as support for the process will continually need to be fostered.

Following analysis of this information, conclusions will be developed and a report prepared and submitted to senior management.

3.2.2 Prepare report incorporating QA amendments

The report

Its contents should be drawn in part from the investigator's conclusions, but should clearly express the opinions and observations of the staff.

The report is likely to outline the following:

- the company's present structure
- the roles and responsibilities of departments and staff
- the company's organisation and methods of control
- difficulties and problems highlighted with particular relevance to divergences from the company's expected structure, roles and organisation
- a particular section related to missing or poorly addressed areas of the International Standard and possible options to address these deficiencies
- a proposed programme for future development.

Structure of the organisation

The structure is likely to develop similarly for all project based companies with a large degree of responsible autonomy at site management level. An example of an extract of a simple structure diagram is shown in Fig 3.2 opposite. Names and information flows would be added to this diagram to provide the detail necessary.

Some typical problems which can be observed from a **detailed** diagram may be:

1.   The geographic division of some individuals' workload may lead to excessive time spent in travelling, thereby reducing their effectiveness and creating paper roles that cannot exist in practice.
2.   Larger than planned quantities of work in strategic areas falling to individuals with dual roles, where competing demands upon their time result either in delay or insufficient attention.
3.   Lack of clarity or understanding of roles means that communication lines in some instances by-pass the formal management structure.
4.   Middle management often have too wide a span of authority with poorly defined responsibility.

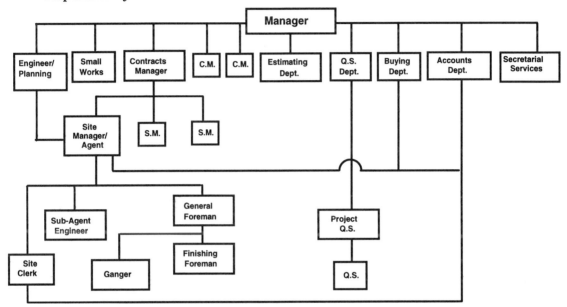

Fig 3.2 Structure diagram

Roles and responsibilities

The following typical points may be highlighted.

Support departments Central support department staff tend to have well defined roles within the department; however their external authority and responsibility is often poorly defined.

Estimating Typical problems may be that:

1.   Staff are subject to a generally heavy though variable workload which means that tenders sometimes suffer in terms of depth of input and thus quality.
2.   Data banks of suppliers and subcontractors are often incomplete and relate heavily to cost (or the ability to return tenders) rather than a balance of cost and quality.

3. Variable or unstructured feedback of information from sites allows the continued use of incorrect data.
4. The experience and availability of site staff leads to wide variations in the extent and quality of their input into tenders.

Quantity surveying Typical problems may be:

1. The amount of time and resources devoted to financial reporting from which projects may gain very little in terms of useful feedback.
2. Problems in reconciling tender allowances from data provided.
3. The time consumed in dealing with large numbers of subcontractors.

Personnel department Typical problems may be:

1. Deficiencies in their records of individuals' past experience and the absence of proof of qualifications.
2. Lack of access to records for site based management.
3. Incompatibility of knowledge of interviewer with the skills of interviewee leads to inappropriate appointments.

Site based personnel Particular roles below that of site agent will be largely at the discretion of the site manager. However there is often confusion at site level when they attempt to define the exact extent of their roles and responsibilities.

Typical problems may be:

1. That because of this lack of definition scope exists for work to be duplicated, poorly addressed or even missed altogether.
2. Titles and roles do not always match, leading to problems when staff are transferred.
3. The support necessary for individuals carrying out the same roles will vary considerably.
4. Individuals with a professional or academic base will express differing needs to those with a practical base and perceive their roles from widely differing viewpoints.
5. Project teams will in some instances tend to lack balance between experience and expertise.

Organisation

Company manuals tend to grow without regard to their usability and thus the processes and procedures that staff frequently use will often be found to be ad hoc developments of what is actually required. Because of this variability management control is often illusory.

Typical problems for site managers from existing organisational methods and control could be:

1. Shortage of time at commencement of a contract leading to an early use of reactive rather than proactive measures which are often a continuing source of problems demanding further time.
2. Inconsistency in transfer of information to site teams - leading to a slower than necessary understanding of the project.
3. Lack of time for supervision - stemming from:

   • shortage of time at commencement

   • poor information flow both internal and external

   • lack of support increasing workload

   • inappropriate proportion of time spent on providing historic cost information.

4. Frequently because of the power of central management, organisations tend to demand information from projects but lack procedures for exchanging information, creating an imbalance which may lead to a breakdown in communication.

3.2.3 Report format

It is worth remembering that the people who will be required to read and understand this report are often pressed for time, so make it concise.

As part of this report flow charts can be used as an aid to graphically identify the involvement of both individuals and departments within the process of completing a project. See Fig 3.3 on next page.

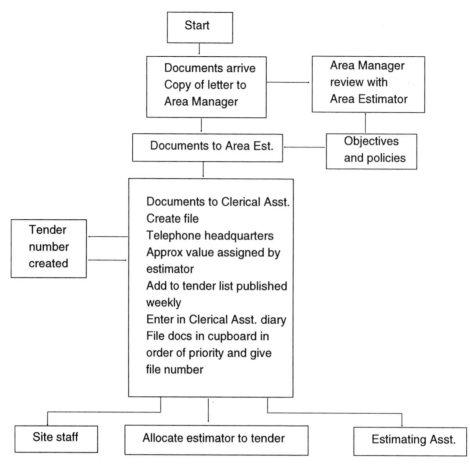

Fig 3.3 Report flowchart extract

Having defined firstly the system required and secondly the company's structure, roles and organisation, actual rather than assumed, it should be possible to identify divergences, contradictions and most importantly missing areas.

From this information a schedule, as previously noted, can be prepared identifying the divergences between the company's current procedures and those necessary for the proposed quality system. Within the report the schedule may offer alternatives as to how these divergences can be answered. See Fig 3.4 on next page.

EXISTING	MISSING	OPTIONS

Contract review
 	
* Pre-construction meeting with client
 	
* Varying appraisal depending upon site staff availability
	* Standardised and resourced appraisal procedures
	
	* Structured and documented arrangement for meetings
		* Formalise existing procedures
		
		* Resources:
		    *appoint new staff
		    *accommodate wider role by modifying existing duties
		    *combination of above

Fig 3.4 Extract of schedule of missing and incomplete procedures, resources and documents related to ISO 9000 requirements

3.2.4 Agree system outline

Management's response to the report in the form of observations and choices should enable an outline of the proposed system to be agreed.

The report and management's response provide a basis for understanding the order of tasks and integration of both individuals and departments within the company's system and would therefore form a starting point for the development of a QA system.

3.3　　Development

The initial flow charts developed for the report together with any management comments should be sent to each department or project and staff asked to comment and expand upon them. From their response a revised flow chart can be prepared and returned to the departments for further comment. Interfaces and processes which are still unclear or contradictory can then resolved by joint discussion, normally by bringing dissenting parties face to face.

Rudyard Kipling wrote a story called The Elephant's Child. In it he referred to six stalwart serving men who served him well and true. He called them What, Why, When, How, Where and Who. A good quality system, in our opinion, must provide these six servants in order for the system to serve the company well:

- defining what is to be done
- explaining why it is done
- establishing when it is to be done
- controlling how it is to be done
- showing where it is done

- identifying who is to do it.

The system thus developed should contain not only the quality related procedures but also dovetail into existing systems of administration, accountability and control. The principle is to set down the system within a main reference document that also has all other company documents referred and subsidiary to it. One solution to achieving this is discussed in detail over the next five chapters.

3.4   Testing

Sections of the system should be introduced progressively on a selective basis. If problems then occur they can be amended to remove the ambiguity or unnecessary item or add any missing points. The system will thereby become refined in use before actually being completed. Upon completion of all stages the documents are brought together combining the requirements of the International Standard with references to the company's own procedures. This will eventually form the principle part of the QA Manual referred to in the next chapter.

Having completed the system it will be necessary to get approval of the whole and to agree a progressive programme of familiarisation and training in the use of the new procedures. The means of maintaining the system should be defined within the system itself. However it must be accepted that the system through use, particularly in its reporting and auditing procedures, will undergo continuous amendment and improvement which is an integral part of QA and looked for by third party auditors.

From the commencement of this process you must keep in mind that the system you are producing will, if it is properly used, be subject to continual improvement. See Fig 3.5.

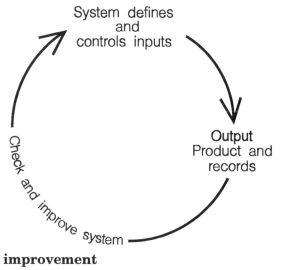

Fig 3.5 Continuous improvement

The use of audits, discussed in Chapter 9, must focus on the procedures and their effectiveness. Effectiveness should be judged, in this context, by the achievement of quality. If quality is consistently achieved without adherence to particular procedures, then the procedures may be unnecessary or incorrect. The aim is not to produce a paper chase but procedures that ensure that quality is achieved.

We believe the system outlined in the following chapters provides a basis to achieve this. However it is worth emphasising that the company's staff should feel involved in the whole process described above and feel of the system 'they did it all themselves'. What we would describe as 'the sense of property' thereby generated will ease the harnessing of their support and help ensure a positive attitude in applying the system.

Remember before designing your system not only what your aims are but also what you must seek to avoid.

Your system should not:

1. inhibit positive innovation
2. feel an imposition
3. be difficult to use

Your system should:

1. assist real tasks not dictate
   superfluous detail
2. make good sense
3. be theirs

Solutions to the points noted are many and various but the remainder of this book provides one solution and some of the reasons for its development.

The next chapter describes the QA Manual and its method of operation.

Relevant accredited Certification Bodies include:

BSI Quality Assurance
PO Box 375
Milton Keynes MK14 6LL

LRQA Ltd
Hiramford
Middlemarch Office Village
Coventry CV 4FJ

NICQA Ltd
26-28 Stuart Street
Luton LU1 2SW

THE SYSTEM DEVELOPED - ITS STRUCTURE AND PHILOSOPHY

4.0 Structure of the system developed

Having talked about QA systems in general we shall now go on to consider the system developed from our own work and its method of operation. This we believe could form a framework for the development of your own system.

Fig 4.1 on page 43 shows the index for a QA Manual illustrating our solution highlighting the necessary component parts. Note that the Manual is divided into five main sections with subdivisions where appropriate.

4.0.1 The main sections
Section 1 The director's statement

Clause 4.1 requires that executive management must define in the form of a written statement its policy, objectives and commitment to quality. This statement must be published throughout the organisation and be seen as supported by management. Clients should also be made aware of this policy.

Policy may be described as a broad statement of the company's organisational goals. Objectives are the means by which the policy may be implemented and should be perceived both as achievable and measurable. Commitment means that executive management is fully behind the policy and objectives.

The obvious place for the publication of the director's statement is the Quality Manual where it forms the introduction. Typical examples are published by the Chartered Institute of Building, Building Employers Confederation and other organisations but a further example is given in Fig 4.5 on page 51.

Section 2 Control systems for Manual

Clause 4.2 requires that the system be defined in a document and that the document must be available to all personnel who have defined roles and responsibilities within it.

It is therefore necessary to produce and circulate copies of the Manual within the organisation. Clause 4.1.3 requires management to regularly review the system to ensure its continuing effectiveness. It is a necessary virtue of an effective appraisal that improvements will have to be made and thus changes to the documentation will be necessary.

Therefore an effective control of these documents will be required. This control has a number of purposes:

1. To record that everyone who is required to have a copy of the Manual has received one.
2. To ensure that any sundry amendments to the Manual are identifiable and circulated to the holders of the Manuals.
3. To ensure that amendments are approved by management prior to their incorporation.
4. To remove obsolete superseded documents.

For the purpose of item one it is necessary that a person is given responsibility for developing a circulation policy and drafting and maintaining a list of those persons to whom the copies are to be issued.

The purpose of item two requires that the document must be produced identifying individual pages by means of a discrete reference. The circulation of amendments to any pages should also be controlled and recorded. See example below:

QA MANUAL	Notes	SECTION	Company Procedures		
		STAGE	Tender		
		REF	QAM 4	T/1	
		REV		DATE	
		SHEET	1	of	22

REF   QAM 4 indicates that the page is within Section 4 of the QA Manual.
T/1   Shows that it is the first page of the Tender stage.
REV   Allows revisions to be enumerated and the date of their issue entered.
SHEET   Shows that it is part of a sequence of 22 sheets.

The purpose of item three is to avoid documents being amended on an ad hoc basis; therefore a procedure for authorising amendments must be specified.

Section 3 Company organisation chart

Clause 4.1.2 requires that the document defines the organisational structure of the company. It is in the nature of project based organisations that their organisational structure will undergo continual changes as existing projects end and new projects start. In satisfying this part of ISO 9000 it is advisable to consider the organisation in two tiers:

- Tier 1 - the relatively stable central administrative organisation

- Tier 2 - the project based organisation.

The first tier is best developed within this section of the Manual. The second tier is project specific and is best developed within the Quality Plan (see Chapter 8). This can then be related in terms of its interfaces and links back to the company organisation chart. The main advantages of doing this are:

- it avoids constant amendment to the manual

- the charts in each Quality Plan are limited to those relevant to each project's staff.

	QA MANUAL INDEX	
Section		Book Reference
1.	Director's statement	Chapter 4
2.	Control system for Manual	Chapter 4
3.	Company organisation chart	Chapter 4
4.	Company procedures	Chapter 4
	1. Staged procedures	
	1.1 Tender stage - Sheets T/1-T/22	Chapter 5
	1.2 Mobilisation stage - Sheets M/1-M/2	Chapter 6
	1.3 Construction stage - Sheets C/1-C/10	Chapter 7
	2. General procedures	
	2.1 Document control	Chapter 8
	2.2 Enquiry	
	2.3 Ordering	
	2.4 Quality Plan	Chapter 8
	2.5 Material, delivery and handling	
	2.6 Quantity Surveyors	
	2.7 Audit	Chapter 9
	2.8 Recruitment / training	Chapter 10
5.	Supporting documentation	
	1. Forms QA/A - QA/Y	Chapter 11
	2. Works guidance notes	
	3. Job descriptions	

Fig 4.1 QA Manual index

Section 4 Company procedures

ISO 9000 requires the contractor (supplier) to produce a Quality Manual which includes the documented procedures consistent with the standard and the contractor's (supplier's) stated quality policy. This is a description of the company's quality objectives and how they are to be achieved. The Quality Manual is therefore the core of any QA system.

In describing the company's activities the report produced from the investigation was drawn from different groups at different stages of the project. Therefore it seemed appropriate to subdivide the system into these same stages of tender, mobilisation and construction. In analysing activities some were found to be discrete to a particular stage and others spanned a number of these stages. In addition certain tasks needed a lengthy explanation which tended to detract from the main flow of information. Therefore company procedures were divided into those that could be accommodated in staged procedures and the other tasks which spanned stages or needed detailed explanation; the latter were separated and called General Procedures. Both of these will be illustrated in some detail in Chapters 5 to 8.

Section 5 Supporting documentation

Noted within Chapters 5 to 8 are documents referred to on the index as Supporting Documentation.

The first documents are the forms which have the function of expanding the information given and also act as live documents within the system. Completion of the forms is part of the procedures within the system and their storage and inspection constitutes a key element in the maintenance and control of the system.

Secondly there are Works Guidance Notes which are an aid to site management. These highlight standard activities and hold points for typical operations found within project Quality Plans (examples are shown within the Quality Plan Procedures pages 124-125).

Job descriptions are the final documents, although as described in the Quality Plan Procedures, these are standard job descriptions which may be subject to written amendment.

4.1   Philosophy of the system developed

The system developed is applicable to a construction organisation's building operations in accordance with BE EN ISO 9002:1994 with a view to obtaining third party certification.

It is a basic premise of quality systems that to control the quality of work, means must be introduced to control the inputs into that work. It is arguable as to what depth a system needs to go. Construction involves a wide variety of inputs all with a multitude of facets occurring over a series of often remote stages. It is therefore difficult to decide how insignificant or remote, from the final product, a task needs to be before control is no longer required. This system has tended in some respects to take this need to what some will consider to be extremes. For instance does the quality of the tender affect the final quality of the constructed building. Research indicates that it does.

Consider the following management problems of time, information, resources and control which are found to be present in varying degrees at all stages of a project.

Typically:

Stage	Problems			
Tender	Time pressure variable	Information missing	Resources variable	Control O.K.
Mobilisation	Time pressure variable	Information missing	Resources variable	Control difficult
Construction	Time pressure periodically	Information periodically	Resources periodically	Control difficult
Defects	Time O.K.	Information O.K.	Resources short	Control variable

Whilst time pressures during the construction stage are sometimes variable it is common in today's industry for the time constraints to be a critical factor from the commencement of many projects. It is at this commencement point that the failings of the processes leading up to the construction stage start to appear. The resolution of these failings will then rob a frequently overstretched site management team of further time. Attention to detail in preparing tenders and setting up a project ensures a maximum use of available time and resources to provide a vitally important platform for a successful project.

This is not to say that from a quality point of view, if this situation should exist by the construction stage, it is irretrievable. However the methods used in retrieving the situation will often affect the balance of time, cost and quality and the contractor is unfortunately faced with having to sacrifice one or two aspects of this eternal triangle.

Another reason for the system having to be developed in some detail is that the ISO Standard specifically requires you to ascertain, prior to submitting your tender, the client's needs and requirements. A contractor (supplier) should not passively or knowingly accept incomplete or incorrect information on the basis that this is the client's or others' problem. It is recognised that the quality of what is received is often beyond control. However the processes of seeking and identifying what is needed, within reason and within the constraints of the constructor's role, must be provided for within the QA system.

For both these reasons it is necessary to investigate and develop flow charts of the tender and mobilisation stages in some detail.

It will be noted in the next three chapters that frequent meetings are called for. This is not a requirement of the ISO Standard but is a useful management tool. It assists in co-ordinating and ensuring that tasks are addressed at the correct stage and that there is a general appreciation by the team of where they are going. These meetings will often be short and relatively informal. They will provide a forum which enables all those with responsibility to communicate or appreciate problems at the earliest possible time.

4.2   The system developed

The system to be produced must incorporate Rudyard Kipling's six servants i.e. the what, when, who, where, how and why of the company's activities. These may be established as follows.

From the distilled information of the investigation it is possible to define **what** is to be done, **when** it is done and **who** does it. Alternative flows caused by varying project types may be discovered and eventually separate processes, each with varying degrees of communality, will be identified. From these alternatives the **where** will be established.

Strategic points along the flow charts can then be identified and a means of control, by the use of checklists and detailed procedures (**how**), developed. Finally in order to obtain a consistency of roles, standard job descriptions written to interface with the flow charts, notes and procedures previously described, can be developed. The job descriptions emphasise the philosophy of the work to be undertaken rather than prescribe a didactic schedule of tasks. The junior positions require more detailed notes on how to undertake particular job-specific tasks although the majority of these may be covered using forms or general procedures.

The system thus emerging consists of a central flow chart of activities highlighting individual responsibilities supported by explanatory notes (**why**) and forms, these forms thereby acting as an expansion of the flow chart providing agendas or checklists at strategic points. In addition detailed procedures on particular activities can be prepared and referenced either on the flow chart or within the forms.

The system therefore consists of:

1. The central flow charts trace the core activities in some detail.
2. The signposting on the flow charts establishes responsibility for tasks.
3. The explanatory notes on or adjacent to the flow charts provide guidance and expand the detail of the task.
4. The forms referred to in the flow chart or notes provide a deeper insight into the detail of the task, amounting almost to a step by step guide as well as acting as a permanent record.
5. The general procedures describe general tasks in detail to ensure their execution in a uniform manner.
6. The job descriptions explain management's perspective of the general nature of the role the individual is expected to undertake.

Any system must have a core or starting point and whilst much of what is shown on the flow chart or forms may appear to be simplistic, our research has shown that they are frequently the very points which are missed or poorly addressed. We start therefore from the viewpoint of taking nothing for granted.

An extract from Page T/6 (page 63 in this book) of the tender stage of the flow chart is shown in Fig 4.2.

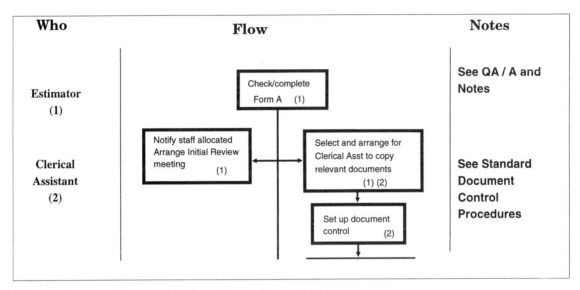

Fig 4.2 Extract from Tender Stage flow chart

The columns are headed Who, Flow and Notes. The Flow column is the core of the system, the boxes describing tasks, generally in the order of their execution, and indicating by means of a numbered key the person or persons responsible for ensuring the tasks are undertaken. The Who column provides a permanently visible responsibility key. The Notes column will normally direct that person to a more detailed explanation either on the facing page or elsewhere in the manual.

For example the relevant facing page for Page T/6 (Fig 4.3 over page) describes the nature of the task in completing any missing sections on Form QA/A designed to set out the basic enquiry information. The forms themselves are listed alphabetically and are kept within the Supporting Documentation section of the Manual. The forms are designed to be used to provide permanent records and as such become part of the company's standard stationery.

> ❑ **Form QA/A (cont'd)**
> The Estimator checks the form and adds any missing information. A copy of the form is to be filed in the 'Records Section' of the Main/Tender/Contract file as part of the document control procedures by the Clerical Assistant. See Document Control Procedures.

Fig 4.3 Notes from page facing T/6

Reference is also made to Document Control procedures. This and other General Procedures are included for complex tasks peripheral to the main flow of activities and are included within Section 4.2 of the Manual.

So within the short extract of the flow chart shown in Fig 4.2:

1. The Estimator has

- checked and/or completed the form which will provide a permanent record of the Tender details (such details must therefore have been considered at this stage)

- selected the documents to be issued to other staff involved

- set up the first meeting for the team. The philosophy of the meeting (which is shown in Fig 4.4) is described later in the flow through Form QA/E which also acts as an agenda.

2. Management has

- defined how keenly the company is viewing the tender at this point in time

- decided what staff are allocated to the tender.

3. The Clerical Assistant has

- filed the form in the records section of the appropriate file

- copied appropriate documents for staff

- set up the Document Control procedures - use of codes, format of files.

> ❑ **Form QA/E Initial Review Meeting**
> The purpose behind the meeting is to establish at an early stage:
>
> 1. The formation of a project team.
> 2. The agreement of a tender timetable.
> 3. An appreciation of the adequacy of the tender information.
> 4. An outline of the method and organisation envisaged.
> 5. The allocation of tasks to members of the team.
>
> The form provides a basic agenda and acts as part of a permanent record of the initial decisions taken. It also ensures that some of the questions required for the Contract Review meeting with the client are established at an early date.

Fig 4.4 Notes for Form QA/E

If you refer to the Form QA/E on page 150 it will indicate how the above philosophy is addressed in a practical manner.

The system is therefore aiming to cover the tasks in great detail whilst keeping them in linked but manageable portions like the layers of a sandwich cake. The conflict between providing forms which act as concise checklists and providing sufficient space for the entry of records has been catered for by allowing the attachment of further sheets which are to be referred to on the forms.

4.3 The operation of the system

Each individual with responsibility is required to have a copy of the QA Manual. It is however possible to issue only the sections on procedures that will apply to that particular individual so that certain individuals will only receive those sections that are relevant to their particular role.

The order and responsibility for all the significant tasks undertaken are followed by means of the flow charts. The points of control are those that have forms to be completed or meetings to be held and recorded. The forms and agendas have the additional function of expanding the areas which management require to be consistently addressed.

The completion of the forms and the additional detail from the meetings provide a permanent record that the tasks have been addressed. Should there be any future problems then the source of the problem should be traceable.

The system therefore:

- provides guidance for the individual
- control for the manager

- assurance for the client.

A more detailed explanation of the operation of this section of the Manual can be found on page 58 which shows the actual guidance notes incorporated within the Manual.

The production of a system heavily dependent upon graphic forms and flow charts requires that such documents should be capable of amendment within the organisation. It is therefore advisable, where possible, to develop all material on a desk-top publishing system. By this means amendments can be easily and quickly incorporated both during development and future use of the system.

The next three chapters set out the system developed in detail, each chapter being prefaced with a brief explanation of the logic which led to the controls and procedures incorporated within each stage.

Chapter 8 looks in detail at procedures for preparing and maintaining Quality Plans and Document Control, which are key elements in a QA system.

Our company's corporate purpose is to provide a quality service to our clients in providing their specified requirements. Only by so doing will we achieve our aims of sustained improvement and profitability.

It is my opinion that the use of this Quality Manual will greatly contribute to these aims.

All personnel within the company are individually responsible for the quality of their work. The procedures contained within this Manual are designed to assist all employees in ensuring that work for which they are responsible meets required standards.

The Quality Manager will monitor and report directly to me on the implementation and effectiveness of this Manual in achieving the above aims.

Signed   (Managing Director)

Date

Fig 4.5 Example of Director's statement

Authors' note

A signature on a Quality Policy statement is not mandatory, it is traceability to the person with executive responsibility for quality that is required.

This page left intentionally blank

THE SYSTEM DEVELOPED - TENDER STAGE

5.0 The points addressed

The system developed for this stage addresses two principle issues:

- the ISO requirements
- the specific problems discovered at the investigation stage.

The clauses of ISO 9002 : 1994 accounted for at this stage were:

4.1 Management Responsibility
4.2 Quality System
4.3 Contract Review
4.5 Document and Data Control
4.6 Purchasing
4.9 Process Control
4.16 Control of Quality Records
4.17 Internal Quality Audits
4.18 Training

Whilst the problems discovered were many and varied the major aspects occurred broadly in four areas:

- time availibility for tenders varied but tended to be limited to 4-6 weeks which was frequently inadequate
- resources were generally stretched and poorly distributed
- information was of variable quality but tended to be incomplete or inaccurate
- control tended to be good because the activities were systematically undertaken and checked but some problems existed.

5.1 Issues considered under the relevant ISO headings

The two principle issues addressed are considered here jointly under the appropriate ISO reference heading. The discussion thereby blends the necessary improvements to existing procedures with the requirements of the International Standard ISO 9000.

5.1.1 Management Responsibility

Management's policy and objectives are defined within the director's statement previously noted in Chapter 4.

The fact that the ISO Standard requires them to be implemented from the top down places a discipline on **all** members of the organisation.

Time and resources were seen as particular problems which a management input could serve to solve. Some early decisions from management were incorporated as part of the system. The Enquiry Review (see diamond on flow chart T/4 page 61) requires the manager to classify the tender into one of four categories; depending upon this category the staffing level is established and the individuals named and allocated to the tender team. *Authors' note - before referring to the specific flow charts examine pages 58 and 58 for an explanation of their methodology.*

Management must recognise the importance of allowing the estimator as much time as possible to maximise the use of resources early. The system provides a discipline which increases the time available on appropriate projects as the categorisation enables an appropriate level of resources to be directed earlier. Some tenders would receive a lower staff input thus freeing resources for projects in a higher category. Flexibility was also required and this was catered for in a controlled way.

This became an enabling process with the flow chart defining the responsibility and role for **all** staff.

Management's role in administration is reflected in the company's training policy. Control is provided for within the system and the responsibility is defined within the Audit procedures.

5.1.2 Quality System

The system is defined and the documents are available to the necessary personnel. The organisational structure at each stage is defined within various documents.

At this stage:

- staff - defined on Form QA/A (page 142)
- roles - are defined at the Initial Review meeting (see Flow Chart T/6 page 63 and Form QA/E page 150)
- processes - are defined by compliance with forms and general procedures.

5.1.3 Contract Review

ISO 9000 requires the contractor to institute a procedure which examines, resolves and documents the process of ensuring that the client's specification is understood. Part of this process is to arrange a meeting with the client's representative. Whilst it is recognised that not all clients would wish to co-operate in such a process, it does provide the optimum solution. The meeting is shown on Flow Chart T/8 (page 65) and the agenda provided by Form QA/G (page 152). In addition processes of inspection and evaluation of information received are provided for, see Form QA/E (page 150).

5.1.4 Document and Data Control

Obtaining and documenting information is a problem at most stages of a project. A further difficulty is ensuring that everyone knows what information has been received, where it is and who has received it.

Document control is introduced at this stage using a modified version of the full Document Control procedures, see Flow Chart T/8 (page 65). Refer to Chapter 8 for more detail.

5.1.5 Purchasing

Time, resources and information normally mitigate against too extensive a procedure at this stage. However it is advisable to do all that is possible to ensure that the tender prices reflect the specified requirements. This is catered for within the Enquiry and Checking procedures on Forms QA/C (pages 145-146) and QA/F (page 151), see Flow Chart T/6 (page 63).

A difficult aspect of ISO 9000 was the requirement to ensure that customer (client) supplied services or products, including information, complied with the appropriate quality standard. Here it was decided not to extend the contractors' responsibility beyond the liability created by the duty of care.

5.1.6 Process Control

The effective use of time and resources is critical at this stage, therefore process control can only receive limited attention. The involvement of available site staff and the planning engineer at this early stage should however enable some basic planning to enable completion of the Programme and Method Review Report on Form QA/B1 (page 143), see Flow Chart T/10 (page 67).

5.1.7 Quality Records

The forms together with the implementation of the Document Control procedures are designed to answer this particular requirement.

5.1.8 Internal Quality Audits

The planned sequence of internal audits is defined within the Audit procedures and will depend to a large extent upon reference to the quality records provided by the system. Audits are thereby simplified by the use of Forms QA/A to QA/K relevant to this stage.

These provide an accessible and complete record which answers a large proportion of the auditor's questions. See Chapter 9 for more details.

5.1.9 Training

The selection of the best available staff is important even at this early stage. Detailed records of staff experience and training ensure that management are provided with the relevant information to aid this process. Two other factors are important. Firstly recruitment procedures must be tightened up. Secondly career structures for staff have to be developed and a system of monitoring them instigated. See Chapter 10 for more details.

❑ **A detailed edited extract of this section of the Manual follows.**

All pages with the QA Manual heading are taken directly from the Manual.

Authors' note

1. *The terminology used within the system is the terminology which is familiar to the industry and not necessarily that used within ISO 9000 e.g. both subcontractors and suppliers would be referred to as subcontractors in ISO 9000 but are referred to by their industry title in the extracts of the system.*

2. *Whilst for completeness procedures for all categories of tender have been included, for ease of understanding you are advised initially to follow Category A procedures only (T/1 - T/12) and refer to Category D (T/22).*

    Subsequently you may wish to examine the alternative procedures for Categories B and C.

Notes for guidance in use of QA Manual

Tender Stage (QA Manual - Flow Charts Pages T/1 to T/22)

Generally

The flow chart plots the progress of a tender from receipt to submittal. The left hand column indicates and numbers the parties involved in the section of work covered by each page. The centre column traces the activities which are to be carried out and are numbered to indicate who does them. The right hand column refers you to forms, copies of which are to be found in Section 5.1 Forms, or are simply notes on how to proceed.

An example of how this manual should be used is as follows:-

All procedures start at Flow Chart QA Manual Page T/4 - refer now to this page.

The main sequence on the left hand side of the **centre** flow column is the company's procedures for receiving and recording tender documents. On the right hand side of the centre column a copy of the covering letter has gone to the Area Manager which ensures that he is aware of their receipt. *The initial tasks and the persons responsible are shown in the main sequence boxes.*

Closely following this receipt a decision is taken by the Area Manager and Area Estimator at the Enquiry Review concerning the company's interest in the tender and staff availability *(the diamond box indicates a decision point).*

During this stage Form QA/A should be completed by the Clerical Assistant giving all necessary background information on the tender, plus recording the decisions taken at the Enquiry Review.

For this example we are assuming the tender is a traditional one and the decision taken at the Enquiry Review has been to place it in Category B.

This means that because either the company is not sure of its interest or staff are unavailable, it will not receive at this initial stage the full input of resources demanded for Category A procedures.

The continuation of procedures for Category B is on flow chart Page T/14.

From this point the Estimator has responsibility for ensuring that all procedures are adhered to.

All forms noted in the right hand column must be completed.

The forms are read in conjunction with notes on the facing page to the flow chart. This enables completion of the form to a standard that produces a checklist of activities and ultimately a record of information and decisions taken.

If there is insufficient room within a section of the form please use continuation sheets but these must be cross-referenced at that section of the form.

Following Category B procedures through to Page T/16 after the Contract Review meeting with the client there is a diamond box where a further decision must be taken.

As can be seen on Notes to Page T/3 a decision to upgrade a Category B to A must not be made later than 2 weeks from the tender submittal date.

At this point a decision based upon a re-appraisal of interest or staff availability will result in the tender going either to an A or D category.

If the decision is D the documents are given to the Area Estimator and the procedures continue on Page T/22.

If alternatively the decision is to upgrade the tender to A then the procedures continue on Page T/16 until you reach 'Go to A'. This directs you back to Page T/8 (following the decision point - Review Category) where the Category A procedures then take you through to the end of the Tender Stage - Page T/12.

Had the tender been a Traditional Category C, A or D or a Design and Build type alternative routes are also given.

Notes for Page T/4

❑ Form QA/A Enquiry Report

The form is required to be completed in stages.

Clerical Assistant (2) completes as much as possible of the first two sections. Information for the remaining sections will be provided by the Area Estimator following the Enquiry Review. The Enquiry Review identifies the method of tendering, category of tender (this category should be reviewed following the Contract Review meeting) and staff allocated. The process by which the tender proceeds is dependent upon these decisions.

Enquiry Review meeting

At this meeting an initial decision will be taken with regard to the company's input into the tender preparation. The job will be given a Category or Type and the names of staff allocated will be stated by the area manager. This decision may be amended in accordance with the following timescale diagram.

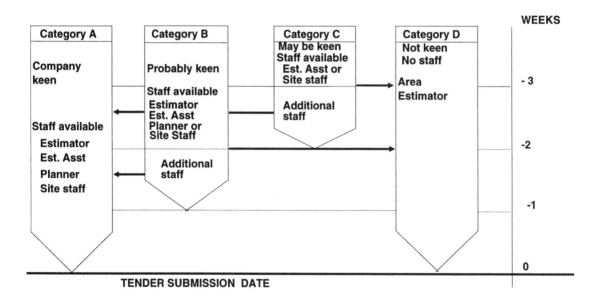

QUALITY ASSURANCE

Who	Flow	Notes

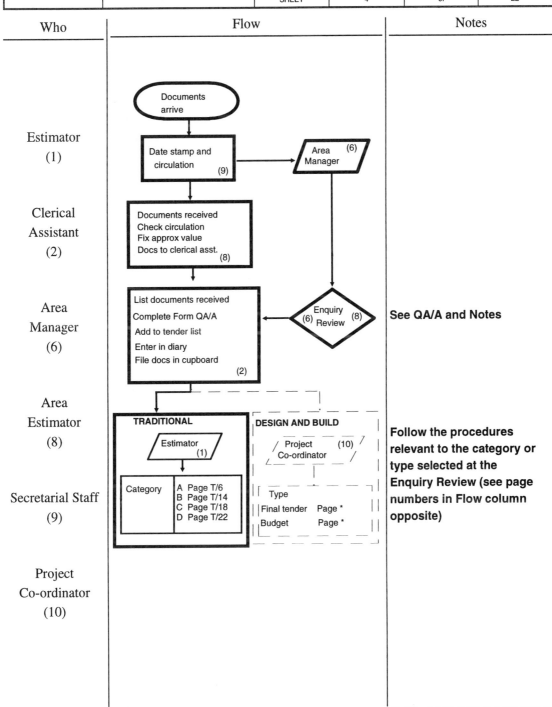

Who column:
Estimator (1)

Clerical Assistant (2)

Area Manager (6)

Area Estimator (8)

Secretarial Staff (9)

Project Co-ordinator (10)

Flow column:
Documents arrive

Date stamp and circulation (9)

Area Manager (6)

Documents received
Check circulation
Fix approx value
Docs to clerical asst. (8)

List documents received
Complete Form QA/A
Add to tender list
Enter in diary
File docs in cupboard (2)

Enquiry Review (6) (8)

TRADITIONAL
Estimator (1)

Category
A Page T/6
B Page T/14
C Page T/18
D Page T/22

DESIGN AND BUILD
Project Co-ordinator (10)
Type
Final tender   Page *
Budget   Page *

Notes column:
See QA/A and Notes

Follow the procedures relevant to the category or type selected at the Enquiry Review (see page numbers in Flow column opposite)

Notes for Page T/6   Traditional Category A Procedure.

❑ Form QA/A (cont'd)

The Estimator checks the form and adds any missing information. A copy of the form is to be filed in the 'Records Section' of the Main/Tender/Contract file as part of the document control procedures by the Clerical Assistant. See Document Control Procedures.

❑ Form QA/C Enquiry Procedures

The purpose of the procedures is to try and ensure that quotations used in tenders are only taken from suppliers and subcontractors that meet our Company's requirements. The form enables the Estimating Assistant/Estimator to list in abreviated form all the quotations sought from both suppliers and subcontractors and to note in the 'CAT' column the criteria by which they were included. New firms included under criterion C (or A in circumstances noted on form) must be checked by the Estimator - see Form QA/F.

Contract Review meeting and site visit The company needs to gain information **from** and provide information **to** the client/client's agent, on its requirements to produce a Quality Assured building. This must be done at the earliest possible stage. Details of the agenda for this meeting will be described later under QA/G. Note: It should be made clear to the client that should this meeting not take place then the company's QA procedures cannot be fully adhered to.

❑ Form QA/D Site Visit Report

The form acts as both a checklist and permanent record of the results of a site visit. The form should be completed by the Estimator or a senior member of the site staff allocated to the project.

❑ Form QA/E Initial Review meeting

The purpose behind the meeting is to establish at an early stage:
1.   The formation of a project team.
2.   The agreement of a tender timetable.
3.   An appreciation of the adequacy of the tender information.
4.   An outline of the method and organisation envisaged.
5.   The allocation of tasks to members of the team.

The form provides a basic agenda and acts as part of a permanent record of the initial decisions taken. It also ensures that some of the questions required for the Contract Review meeting with the client are established at an early date.

❑ Form QA/F Checking Procedures

The form outlines the criteria by which the Estimator can establish a supplier's or subcontractor's conformance. Any divergence from the points listed or new criteria created should be noted. The Estimator should indicate within the 'checked' column of QA/C which of the criteria listed has enabled him to include the company. It is not required at this stage that documentary evidence of conformance be obtained.

Who	Flow	Notes

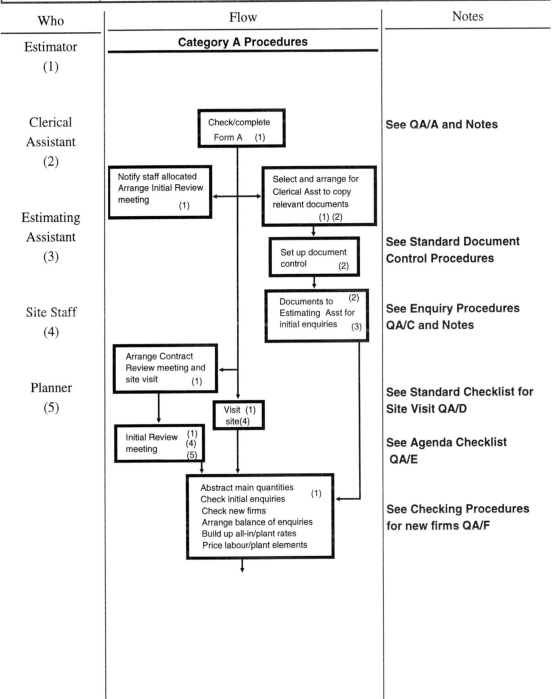

Who

Estimator (1)

Clerical Assistant (2)

Estimating Assistant (3)

Site Staff (4)

Planner (5)

Flow

Category A Procedures

Check/complete Form A (1)

Notify staff allocated Arrange Initial Review meeting (1)

Select and arrange for Clerical Asst to copy relevant documents (1) (2)

Set up document control (2)

Documents to Estimating Asst for initial enquiries (2) (3)

Arrange Contract Review meeting and site visit (1)

Visit (1) site(4)

Initial Review meeting (1) (4) (5)

Abstract main quantities (1)
Check initial enquiries
Check new firms
Arrange balance of enquiries
Build up all-in/plant rates
Price labour/plant elements

Notes

See QA/A and Notes

See Standard Document Control Procedures

See Enquiry Procedures QA/C and Notes

See Standard Checklist for Site Visit QA/D

See Agenda Checklist QA/E

See Checking Procedures for new firms QA/F

Notes for Page T/8   Traditional Category A Procedure.

❑ Form QA/G Contract Review

The form acts as a checklist/agenda and should be supported by brief minutes prepared by the Estimator.

Who	Flow	Notes

Category A Procedures

Estimator (1)

Clerical Assistant (2)

Estimating Assistant (3)

Site Staff (4)

Planner (5)

Area Manager (6)

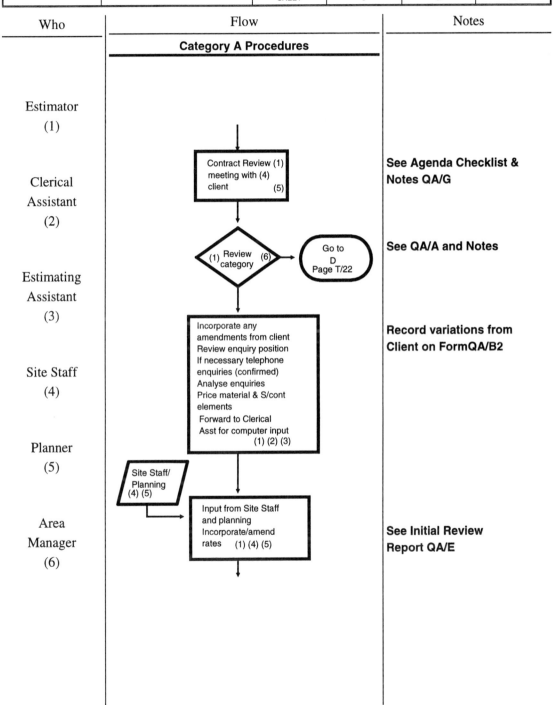

Contract Review (1) meeting with (4) client (5)

See Agenda Checklist & Notes QA/G

(1) Review category (6) — Go to D Page T/22

See QA/A and Notes

Incorporate any amendments from client
Review enquiry position
If necessary telephone enquiries (confirmed)
Analyse enquiries
Price material & S/cont elements
Forward to Clerical Asst for computer input (1) (2) (3)

Record variations from Client on FormQA/B2

Site Staff/ Planning (4) (5)

Input from Site Staff and planning
Incorporate/amend rates (1) (4) (5)

See Initial Review Report QA/E

TENDER STAGE

65

Notes for Page T/10   Traditional Category A Procedure.

❑ **Form QA/B1 Programme and Method Review Report**

The meeting should allow the Estimator to take account of the work of the Site Staff and Planner and provide a check of major elements of his tender. The form again acts as a checklist and records their discussion as an aid to the Area Manager prior to the Tender Review meeting.

❑ **Form QA/B2 Estimator's Report**

The estimator is to note any amendments by the client to the original documents notified either verbally or in writing. Verbal notifications are to be confirmed in writing.

A second estimator is required to carry out a superficial analysis of the priced BQ and will note any high or low rates in the first column. If 'own rate' what part of the rate is high or low is noted in column 2 and a brief comment made in column 4. The Estimator receives the report back from the second estimator and notes in the final column whether the tender has been amended. As with Form QA/B1 the information is provided as an aid to the area manager in the Tender Review meeting. It will also allow the discovery of any obvious errors that might otherwise be missed.

❑ **Form QA/H Tender Review Record**

The form provides an outline agenda and record of decisions taken at the Tender Review meeting.

Who	Flow	Notes

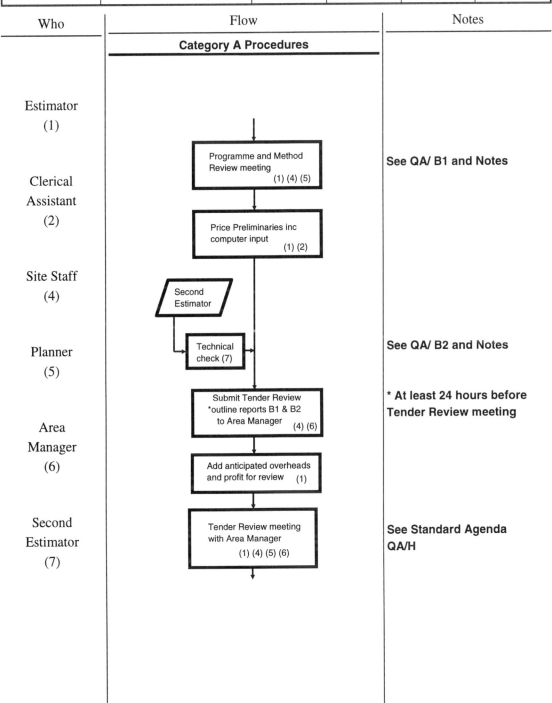

Category A Procedures

Estimator (1)

Programme and Method Review meeting (1) (4) (5)

See QA/ B1 and Notes

Clerical Assistant (2)

Price Preliminaries inc computer input (1) (2)

Site Staff (4)

Second Estimator

Planner (5)

Technical check (7)

See QA/ B2 and Notes

Submit Tender Review *outline reports B1 & B2 to Area Manager (4) (6)

* At least 24 hours before Tender Review meeting

Area Manager (6)

Add anticipated overheads and profit for review (1)

Second Estimator (7)

Tender Review meeting with Area Manager (1) (4) (5) (6)

See Standard Agenda QA/H

Notes for Page T/12   Traditional Category A Procedure.

There are no notes for this page.

Who	Flow	Notes

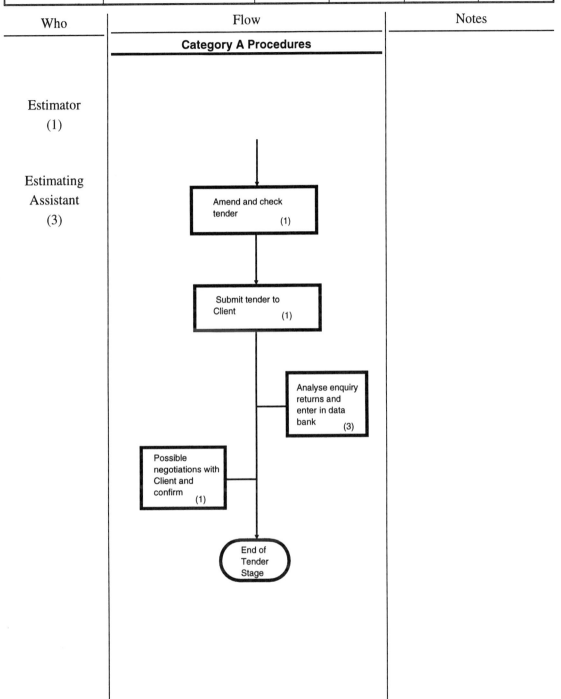

Category A Procedures

Estimator

(1)

Estimating
Assistant

(3)

Amend and check tender (1)

Submit tender to Client (1)

Analyse enquiry returns and enter in data bank (3)

Possible negotiations with Client and confirm (1)

End of Tender Stage

Notes for Page T/14   Traditional Category B Procedure.

❑ Form QA/A (cont'd)

The Estimator checks the form and adds any missing information. A copy of the form is to be filed in the 'Records Section' of the Main/Tender/Contract file as part of the document control procedures by the Clerical Assistant. See Document Control Procedures.

❑ Form QA/C Enquiry Procedures

The purpose of the procedures is to try and ensure that quotations used in tenders are only taken from suppliers and subcontractors that meet our company's requirements. The form enables the Estimating Assistant/Estimator to list in abbreviated form all the quotations sought from both suppliers and subcontractors and to note in the 'CAT' column the criteria by which they were included. New firms included under criterion C (or A in circumstances noted on form) must be checked by the Estimator - see Form QA/F.

Contract Review meeting and site visit The company needs to gain information **from** and provide information **to** the client/client's agent, on its requirements to produce a Quality Assured building. This must be done at the earliest possible stage. Details of the agenda for this meeting will be described later under QA/G. Note: It should be made clear to the client that should this meeting not take place then the company's QA procedures cannot be fully adhered to.

❑ Form QA/D Site Visit Report

The form acts as both a checklist and permanent record of the results of a site visit. The form should be completed by the Estimator or a senior member of the site staff allocated to the project.

❑ Form QA/E Initial Review meeting

The purpose behind the meeting is to establish at an early stage:

1. The formation of a project team.
2. The agreement of a tender timetable.
3. An appreciation of the adequacy of the tender information.
4. An outline of the method and organisation envisaged.
5. The allocation of tasks to members of the team.

The form provides a basic agenda and acts as part of a permanent record of the initial decisions taken. It also ensures that some of the questions required for the Contract Review meeting with the client are established at an early date.

❑ Form QA/F Checking Procedures

The form outlines the criteria by which the Estimator can establish a suppliers' or subcontractors' conformance. Any divergence from the points listed or new criteria created should be noted. The Estimator should indicate within the 'checked' column of QA/C which of the criteria listed has enabled him to include the company. It is not required at this stage that documentary evidence of conformance be obtained.

Who	Flow	Notes

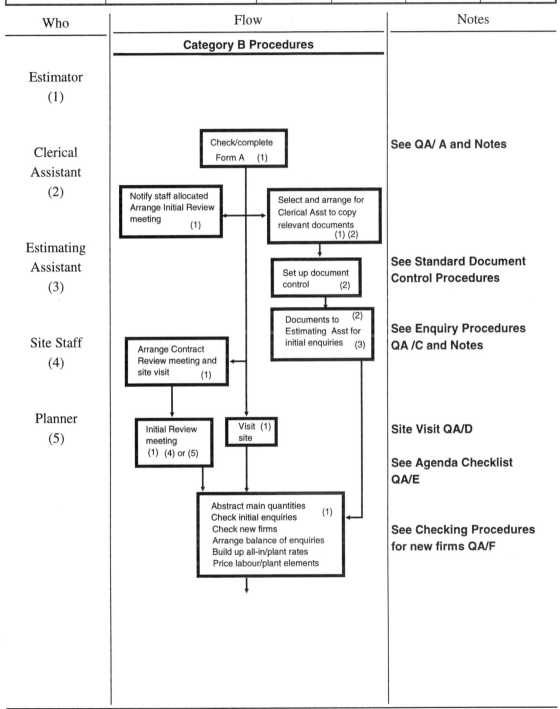

Category B Procedures

Estimator (1)

Clerical Assistant (2)

Estimating Assistant (3)

Site Staff (4)

Planner (5)

Check/complete Form A (1)

See QA/ A and Notes

Notify staff allocated Arrange Initial Review meeting (1)

Select and arrange for Clerical Asst to copy relevant documents (1) (2)

Set up document control (2)

See Standard Document Control Procedures

Documents to Estimating Asst for initial enquiries (2) (3)

See Enquiry Procedures QA /C and Notes

Arrange Contract Review meeting and site visit (1)

Initial Review meeting (1) (4) or (5)

Visit (1) site

Site Visit QA/D

See Agenda Checklist QA/E

Abstract main quantities Check initial enquiries Check new firms Arrange balance of enquiries Build up all-in/plant rates Price labour/plant elements (1)

See Checking Procedures for new firms QA/F

Notes for Page T/16   Traditional Category B Procedure.

❏ Form QA/G Contract Review

The form acts as a checklist/agenda and should be supported by brief minutes prepared by the Estimator.

If Category A.

Second Review meeting follows the pattern of the Initial Review and is intended to:

1. Establish the enlarged project team.
2. Critically review the original timetable.
3. Review the increased information available in terms of of its adequacy.
4. Gain agreement of the larger team to the method and organisation proposals.
5. Re-allocate tasks to the enlarged team.

Who	Flow	Notes

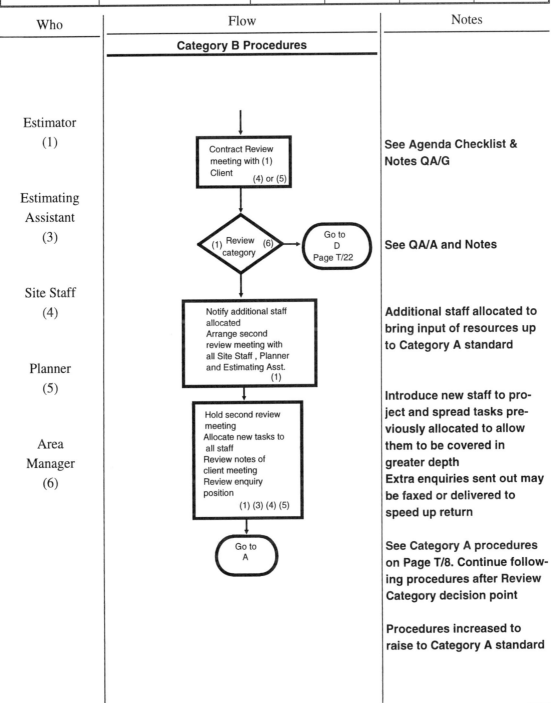

Category B Procedures

Who:

Estimator (1)

Estimating Assistant (3)

Site Staff (4)

Planner (5)

Area Manager (6)

Flow:

Contract Review meeting with (1) Client (4) or (5)

(1) Review category (6) → Go to D Page T/22

Notify additional staff allocated
Arrange second review meeting with all Site Staff , Planner and Estimating Asst. (1)

Hold second review meeting
Allocate new tasks to all staff
Review notes of client meeting
Review enquiry position (1) (3) (4) (5)

Go to A

Notes:

See Agenda Checklist & Notes QA/G

See QA/A and Notes

Additional staff allocated to bring input of resources up to Category A standard

Introduce new staff to project and spread tasks previously allocated to allow them to be covered in greater depth
Extra enquiries sent out may be faxed or delivered to speed up return

See Category A procedures on Page T/8. Continue following procedures after Review Category decision point

Procedures increased to raise to Category A standard

Notes for Page T/18   Traditional Category C Procedure.

❑ Form QA/A (cont'd)

The Estimator checks the form and adds any missing information. A copy of the form is to be filed in the 'Records Section' of the Main/Tender/Contract file as part of the document control procedures by the Clerical Assistant. See Document Control Procedures.

❑ Form QA/D Site Visit Report

The form acts as both a checklist and permanent record of the results of a site visit. The form should be completed by the Estimator or a senior member of the site staff allocated to the project.

Who	Flow	Notes

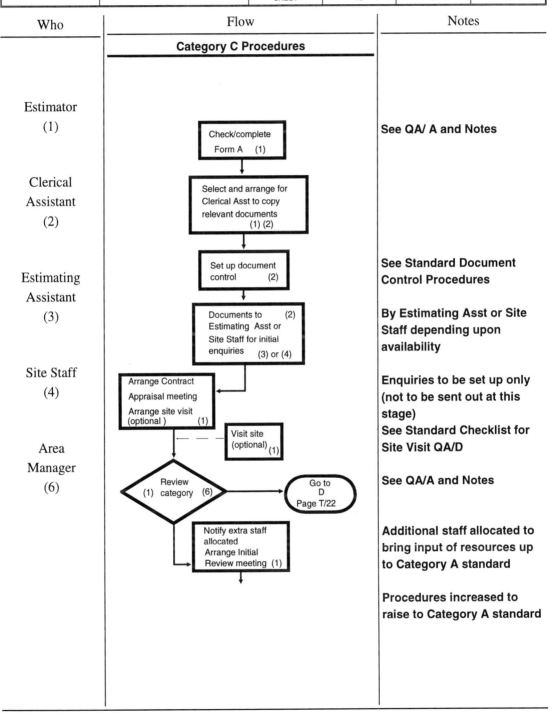

Category C Procedures

Who:

Estimator (1)

Clerical Assistant (2)

Estimating Assistant (3)

Site Staff (4)

Area Manager (6)

Flow:

- Check/complete Form A (1)
- Select and arrange for Clerical Asst to copy relevant documents (1) (2)
- Set up document control (2)
- Documents to Estimating Asst or Site Staff for initial enquiries (2) (3) or (4)
- Arrange Contract Appraisal meeting / Arrange site visit (optional) (1)
- Visit site (optional) (1)
- Review category (1) (6) → Go to D Page T/22
- Notify extra staff allocated / Arrange Initial Review meeting (1)

Notes:

See QA/ A and Notes

See Standard Document Control Procedures

By Estimating Asst or Site Staff depending upon availability

Enquiries to be set up only (not to be sent out at this stage)
See Standard Checklist for Site Visit QA/D

See QA/A and Notes

Additional staff allocated to bring input of resources up to Category A standard

Procedures increased to raise to Category A standard

Notes for Page T/20   Traditional Category C Procedures.

❏ Form QA/D Site Visit Report

The form acts as both a checklist and permanent record of the results of a site visit. The form should be completed by the Estimator or a senior member of the site staff allocated to the project.

❏ Form QA/C Enquiry Procedures

The purpose of the procedures is to try and ensure that quotations used in tenders are only taken from suppliers and subcontractors that meet our company's requirements. The form enables the Estimating Assistant/Estimator to list in abreviated form all the quotations sought from both suppliers and subcontractors and to note in the 'CAT' column the criteria by which they were included. New firms included under criterion C (or A in circumstances noted on form) must be checked by the Estimator - see Form QA/F.

Contract Review meeting and site visit The company needs to gain information **from** and provide information **to** the client/client's agent, on its requirements to produce a Quality Assured building. This must be done at the earliest possible stage. Details of the agenda for this meeting will be described later under QA/G. Note: It should be made clear to the client that should this meeting not take place then the company's QA procedures cannot be fully adhered to.

❏ Form QA/E Initial Review meeting

The purpose behind the meeting is to establish at an early stage:
1. The formation of a project team.
2. The agreement of a tender timetable.
3. An appreciation of the adequacy of the tender information.
4. An outline of the method and organisation envisaged.
5. The allocation of tasks to members of the team.

The form provides a basic agenda and acts as part of a permanent record of the initial decisions taken. It also ensures that some of the questions required for the Contract Review meeting with the client are established at an early date.

❏ Form QA/F Checking Procedures

The form is a checklist only which outlines the means by which the Estimator can establish conformance. Any divergence from it or new criteria created should be noted. The Estimator should indicate within the 'checked' column of QA/C which of the criteria listed has enabled him to include the company. He is not required at this stage to obtain documentary evidence of conformance.

❏ Form QA/G Contract Review

The form acts as a checklist/agenda and should be supported by brief minutes prepared by the Estimator.

Who	Flow	Notes

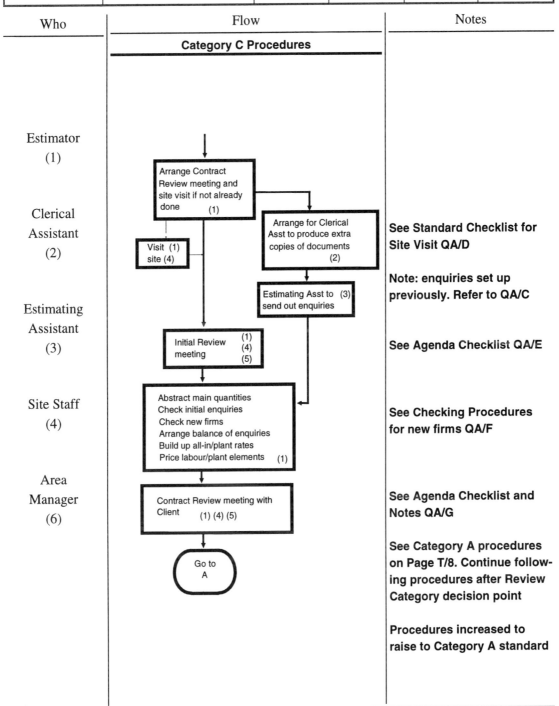

Category C Procedures

Estimator (1)

Clerical Assistant (2)

Estimating Assistant (3)

Site Staff (4)

Area Manager (6)

Arrange Contract Review meeting and site visit if not already done (1)

Arrange for Clerical Asst to produce extra copies of documents (2)

Visit (1) site (4)

Estimating Asst to (3) send out enquiries

Initial Review meeting (1) (4) (5)

Abstract main quantities
Check initial enquiries
Check new firms
Arrange balance of enquiries
Build up all-in/plant rates
Price labour/plant elements (1)

Contract Review meeting with Client (1) (4) (5)

Go to A

See Standard Checklist for Site Visit QA/D

Note: enquiries set up previously. Refer to QA/C

See Agenda Checklist QA/E

See Checking Procedures for new firms QA/F

See Agenda Checklist and Notes QA/G

See Category A procedures on Page T/8. Continue following procedures after Review Category decision point

Procedures increased to raise to Category A standard

Notes for Page T/22   Traditional Category D Procedures.

❑ **Form QA/A (cont'd)**

The Area Estimator checks the form and adds any missing information (e.g. Revised Category and date). A copy of the form is to be filed in the 'Records Section' of the Main/Tender/Contract file as part of the document control procedures by the Clerical Assistant. See Document Control Procedures.

Who	Flow	Notes

Category D Procedures

Area
Estimator
(8)

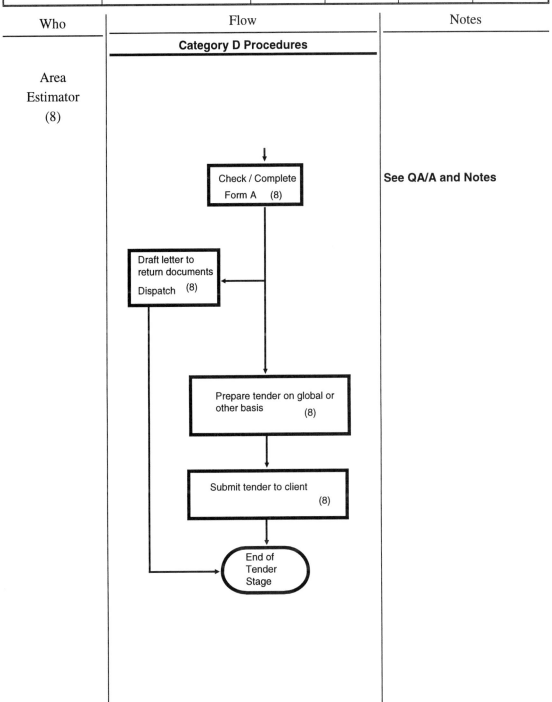

Check / Complete
Form A   (8)

Draft letter to
return documents

Dispatch   (8)

Prepare tender on global or
other basis   (8)

Submit tender to client
(8)

End of
Tender
Stage

See QA/A and Notes

This page left intentionally blank

THE SYSTEM DEVELOPED - MOBILISATION STAGE

6.0   The points addressed

The Mobilisation Stage is the period between the acceptance of a contractor's tender to the start of construction on site. This period varies considerably in length but is a vitally important timespan where available resources must be fully utilised.

As with the previous chapter the system developed for this stage addresses the same two principle issues:

- the ISO 9000 requirements
- the specific problems discovered at the investigation stage.

There is of course a broadening of the relevant ISO 9000 requirements by this stage although the particular clauses addressed remain as the tender stage:

4.1   Management Responsibility
4.2   Quality System
4.3   Contract Review
4.5   Document and Data Control
4.6   Purchasing
4.9   Process Control
4.16   Control of Quality Records
4.17   Internal Quality Audits
4.18   Training

Again the principle problems discovered at this stage and noted in the initial report occurred in the same four broad areas:

- time available for this stage could be variable but was frequently a major concern
- resources were often slow in being released from previous commitments and unbalanced in terms of skills and experience
- information tended to arrive in large and difficult to digest batches to individuals who were often unfamiliar with the project
- control was difficult in this transitional period. The site manager was often tied down with the early procurement and planning of the project. This was exacerbated by having to deal with a team and clients that were often relative strangers.

6.1 Issues considered under the relevant ISO 9000 headings

6.1.1 Management responsibility

The emphasis on management here is to:

- devolve authority and responsibility
- provide resources, including information.

The early appointment of the site manager is seen as essential if the constraints of time and resources are to be minimised (diamond on Flow Chart M/2 page 85).

Staffing new projects at relatively short notice is often a compromise between the ideal and the possible.

The Pre-commencement Handover meeting establishes a formal link between the tender and production teams (Flow Chart M/2 page 85 and Agenda Form QA/L page 155). This helps ensure that the records generated in the tender stage are discussed to facilitate continuity and a rapid assimilation of key information.

The role of management falls largely to the site manager. It is important that the site manager further devolves this role, in a formal and structured way, to the initial project team. In this way whatever resources are available are used to maximum effect (Form QA/M pages 156 - 157).

6.1.2 Quality System

Again at this stage the system is defined, relying more on the detail contained on Forms and General Procedures than on the Flow Chart which is restricted to detailing the principal activities only.

As with the tender stage, staff and their roles are defined, albeit initially on a temporary basis. Processes are again controlled by Forms and General Procedures.

6.1.3 Contract Review

Further information becomes available and has to be examined. In establishing a relationship with the client's representatives it is important to understand their intentions in terms of the provision of available and future information. This is provided by the Pre-commencement meeting with the client (Flow Chart M/2 page 85 and agenda on Form QA/N page 158).

This is intended to build upon the previous pre-tender meeting, assuming it has occurred, and to provide a platform for regular meetings in the future. Early advice on the future supply of information will aid the planning and control of activities at this stage.

6.1.4 Document and Data Control

Continuity is ensured by linking the procedures applicable at this stage to those of the preceding stage. This is done by developing those procedures by increasing the sophistication of the filing system.

6.1.5 Purchasing

This is clearly more important at this stage and reference is made to Enquiry, Ordering and Quality Plan procedures all of which are designed to ensure compliance with the specified requirements. The Detailed Quality Plan will, if available, provide information of any tests or inspections which may be necessary. Orders sent to subcontractors or suppliers must provide these details and require compliance. However a general clause to require their compliance without specifically detailing the tests etc. is included within the standard enquiry form. For more detail refer to Chapter 8.

6.1.6 Process Control

An early start must be made on planning and the development of any written method statements for work occurring early in the programme or with long lead-in periods.

The control of the production process is best established via the Quality Plans. Refer to Chapter 8 for a description of this procedure.

6.1.7 Control of Quality Records

Again the forms and records of meetings will provide evidence of compliance with the Quality System.

6.1.8 Internal Quality Audits

Compliance will largely be judged against the records provided. However it is also necessary at this stage to inspect the Quality Plans to ensure that they have been prepared in compliance with the procedures. Compliance with other General Procedures will normally be established by a planned programme of checks.

6.1.9 Training

Effective use of available staff can only be enhanced by the institution of a training programme. Whilst this programme is primarily directed towards the needs of the company it should also consider the aspirations of the individual. The procedures also provide for an accessible and comprehensive system of personnel records for management at all levels. Refer to Recruitment/Training procedures in Chapter 10.

❑ **A detailed edited extract of this section of the Manual follows.**

All pages with the QA Manual heading are taken directly from the Manual.

Notes for Page M/2   **Traditional Tender**

❑ Form QA/L Handover Meeting Record

The forms act as a checklist and record of the handover from the tender team to the initial site management team. The completed forms handed over provide basic information which together with the discussion should provide a quick insight into the project. Copies of the information to be handed over should be made available to the site manager at least 24 hours before the meeting. Minutes of the meeting should be taken by the site manager and filed together with the form in the 'Records Section' of the Project Filing System.

❑ Form QA/M Project Team Duties Review

The mobilisation period is extremely influential and should not be underrated in terms of the project's future success. The maximum amount of preparatory work possible should be undertaken during this period. The form provides a checklist and requires both a timescale to be set and the allocation of staff duties to be made at an early stage.

Refer also to Document Control, Enquiry and Quality Plan Procedures.

The form recognises the importance of the information provided and requires its early critical examination using Form QA/V.

The early procurement of proper resources is also important to the future success of the project and should be controlled at an early stage.

❑ Form QA/ N Contract Review (Stage 2)

It is important that the company establishes good communication links at an early stage. Information has to be passed in both directions and a knowledge of the procedures of both sides helps this process.

The availability and understanding of information provided cannot commence too soon. The form again provides a checklist and can be used for brief notes. However, minutes should be taken by the company and filed, together with this form, in the Records Section of the Project Filing System.

❑ Form QA/P Pre-commencement Co-ordination Meeting

This meeting brings together the larger team now established. The duties assigned at the earlier stage are examined and reviewed in terms of progress. Duties which are incomplete may be re-allocated at this stage to involve the wider team.

A critical examination of information requirements and the quality of that provided needs to take place in order to:

1. Allocate further duties.
2. Decide whether an effective start is possible.

A review of the progress made in obtaining resources also takes place.

The site manager should make all the staff aware of their position in the team and the duties they are expected to undertake.

		SECTION	Company Procedures	
QA MANUAL	**Flow Chart**	STAGE	Mobilisation	
		REF	QAM 4	M/2
		REV		DATE
		SHEET	2	of	2

Who	Flow	Notes

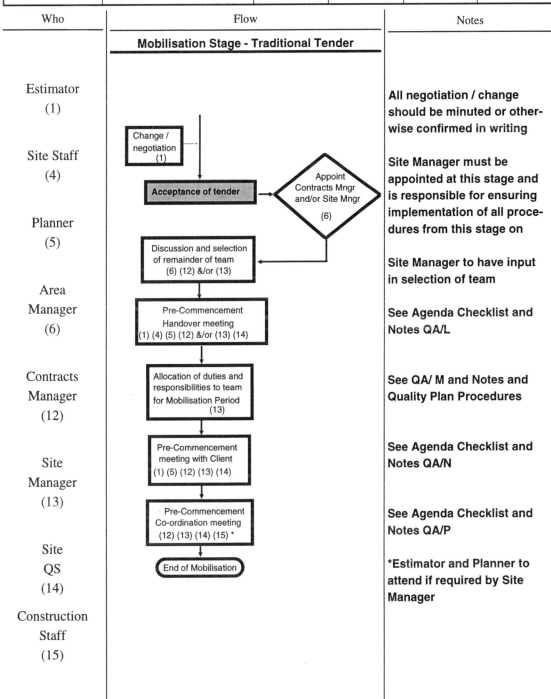

Mobilisation Stage - Traditional Tender

Who column:
Estimator (1)

Site Staff (4)

Planner (5)

Area Manager (6)

Contracts Manager (12)

Site Manager (13)

Site QS (14)

Construction Staff (15)

Flow:
Change / negotiation (1)

Acceptance of tender → Appoint Contracts Mngr and/or Site Mngr (6)

Discussion and selection of remainder of team (6) (12) &/or (13)

Pre-Commencement Handover meeting (1) (4) (5) (12) &/or (13) (14)

Allocation of duties and responsibilities to team for Mobilisation Period (13)

Pre-Commencement meeting with Client (1) (5) (12) (13) (14)

Pre-Commencement Co-ordination meeting (12) (13) (14) (15) *

End of Mobilisation

Notes:
All negotiation / change should be minuted or otherwise confirmed in writing

Site Manager must be appointed at this stage and is responsible for ensuring implementation of all procedures from this stage on

Site Manager to have input in selection of team

See Agenda Checklist and Notes QA/L

See QA/ M and Notes and Quality Plan Procedures

See Agenda Checklist and Notes QA/N

See Agenda Checklist and Notes QA/P

*Estimator and Planner to attend if required by Site Manager

This page left intentionally blank

7

THE SYSTEM DEVELOPED - CONSTRUCTION STAGE

7.0   The points addressed

By this stage the system developed had to account for all the requirements of ISO 9000 noted in Chapter 2 and further aspects of the ISO requirements previously listed in Chapters 5 and 6 as follows:

4.1   Management responsibility
4.2   Quality System
4.3   Contract Review
4.5   Document and Data Control
4.6   Purchasing including Customer (Client) Supplied items (4.7) and Traceability (4.8)
4.9   Process Control
4.10   Inspection and Testing including Equipment Tests (4.11), Inspection/Test status (4.12), Non-conformance Control (4.13), Corrective and Preventive Action (4.14) and the use of Statistical Techniques (4.20)
4.15   Handling and Storage etc.
4.16   Control of Quality Records
4.17   Internal Quality Audits
4.18   Training

In looking at this final stage the problems are again considered in the same four areas of time, resources, information and control:

- time was found to be an increasingly common problem as construction periods on new projects were continually being squeezed. Obviously within the construction period the pressure of time was variable, being linked largely to the performance and availability of resources and the consistency of information flow

- resources were extremely variable in quality and experience resulting in problems arising periodically with little or no notice preceding their occurrence

- information flow was often frequently poorly co-ordinated with production needs. The quality of information was often poor and the discovery of its shortcomings too frequently occurred after the information had been passed on for use by others

- control by management at all levels was found to be difficult. Head office management found difficulty in arriving at a balance between involvement and interference at project level. Site based management's control was often made more difficult if they were spending too much time feeding what was often regarded as superfluous information to senior management. Other problems at project level concerned the number, variety and quality of resources which the site manager was required to manage.

7.1 Issues considered under the relevant ISO headings

7.1.1 Management responsibility

At this stage management is occurring both at different levels and different locations. The system must provide for senior management's role in defining, controlling and administering the system. The system is proactive in setting down the procedures that must be followed to ensure quality. It must also provide for management to be reactive in terms of developing and improving the system on the basis of the information it receives. The site manager's role is to complete the definition of the system through the production of the project specific Quality Plans. Forms and procedures are provided to aid this process, see Chapter 8.

It is recognised that the system should prevent the link between the site manager and area manager/office from becoming too impersonal and from being a one-way flow of information. The Quality Plans set down the management structure at project level, the choice of structure and individual roles. The scope and methods of control within the Quality Plan are left largely to the discretion of the site manager. The aim is to ensure that authority and responsibility for achieving quality at this level continues to rest with site management. Senior management will monitor the project's performance via Monthly Reports (Form QA/T pages 168 - 171) and meetings. The Quality Manager will inspect the Quality Plans and records to ensure conformance.

7.1.2 Quality System

This stage defines the remainder of the system at project level through the flow chart, forms and general procedures. However the majority of the system is project specific and created by the site manager through the development of the Quality Plans. Through this the balance of roles and responsibilities are defined and the processes controlled by again using forms and procedures.

7.1.3 Contract Review

The platform of regular meetings broadens to encompass particular subcontractors and suppliers to ensure that the specified requirements are understood by all parties.

7.1.4 Document and Data Control

Responsibility for ensuring the continuity of the procedures developed for the previous stage rests with the site manager.

7.1.5 Purchasing

Again Enquiry, Ordering and Quality Plan procedures apply to ensure not only compliance with specified requirements but also to stipulate that documentary evidence of compliance is to be provided where requested. Traceability i.e. records of what went where, e.g. a concrete pours record, would form part of the Quality Plans.

7.1.6 Process Control

Detailed planning and the production of written method statements or instructions should continue through the construction stage. This should be linked to the preparation of Detailed Quality Plans.

7.1.7 Inspection and Testing

The Quality Plan procedures involve the development of an Inspection and Testing Plan which combined with the Handling and Storage procedures and the check on receipt (Form QA/Y page 176) ensure that at all stages of construction compliance is verified. A record and schedule of equipment tests is incorporated in the Detailed Quality Plans, as is a system of instructions and records which enable corrective action in the event of non-conformance. The use of statistical techniques to establish statistically acceptable samples rather than blanket inspections is noted within the Quality Plan procedures. See Chapter 8.

7.1.8 Control of Quality Records

Forms, records of meetings, and returns to head office provide evidence of compliance with the Quality System.

7.1.9 Internal Quality Audits

The auditor (Quality Manager) is not responsible for producing quality work. Therefore every effort has been made with the system to avoid placing responsibility for the quality of work with the auditor (Quality Manager). The auditor, and through the audit function management, have responsibility for ensuring compliance with, and or improvement of, the Quality System. The presence and authority of the auditor within the system is given a low profile. Instead, the returns to be provided by the site manager are relied on heavily to monitor the continuing compliance with the system. This is supplemented by the auditor's wider role of support, advice and inspection described in Chapter 9. Records of non-conformance and recommendations for amendments to the system should be forwarded to executive management.

7.1.10 Training

The training procedures require appraisal of new and existing staff. For longer serving staff this is omitted by use of an experience waiver. Responsibility for training, identification of needs and reporting on progress is defined within these procedures. Refer to Chapter 10.

❑ **A detailed edited extract of this section of the Manual follows.**

All pages with the QA Manual heading are taken directly from the Manual.

Authors' note

The terminology used within the system is the terminology which is familiar to the industry and not necessarily that used within ISO 9000 e.g. both subcontractors and suppliers would be referred to as subcontractors in ISO 9000 but are referred to by their industry title in the extracts of the system.

This page left intentionally blank

Notes for Page C/2

❑ QA/Q Site (co-ordination) meeting timetable

The form is designed to provide the basic information required by all parties. Copies of the form providing dates for proposed meetings should be sent to participating consultants/subcontractors as soon as possible. It is recognised that subcontractors will normally attend a separate meeting. An agenda can be found on forms QA/R and QA/R(s).

❑ QA/U Information request

The form allows for written or typed requests to be submitted. You should try and ensure that your questions are sufficiently detailed and unambiguous. Two copies of the form are forwarded to the architect who should be requested to utilise the reply section to link the response to the request. Requests should be referenced in sequence, dated and the date the information is required by entered. The persons authorised to request information will be established by their job descriptions shown in the Quality Plan - see Quality Plan Procedures.

❑ QA/V Information receipt/appraisal

It is important that all information received is recorded (see Document Control Procedures) and appraised prior to filing or forwarding onward for use by others. The aim is to ensure that the information is both correct, as best can be judged, and complete at as early a point in time as possible. This also avoids passing on unknowingly incorrect or incomplete information to those who must use it. Again duties and responsibility for appraising information will be established within the Quality Plans - see Quality Plan Procedures.

❑ QA/W

Part of the Quality Plan Procedures is to establish procedures at site level for the inspection and testing of work. The form has been designed for use as part of these procedures in inspecting our own work or our subcontractor's work. Hold points will have been defined within the Quality Plan and marked on the detailed programme. Waivers are the acceptance of work which does not conform and can only be given with the approval of the site manager.

Who	Flow	Notes

On Commencement

Who:

Site
Manager
(13)

Site QS
(14)

Construction
Staff
(15)

Site
Clerk
(16)

Flow:

Check architects setting out
detail drawings

Complete survey / initial setting out

Complete accommodation / storage
and services as far as possible

Establish material receipt / handling
/storage arrangements

Arrange timetable for Site (Co-ordination)
meetings with architect / subcontractors

Appoint safety supervisor
and first aider/s

Continue to develop detailed Quality Plans
and highlight significant items on the
short-term programme and items with long
lead in on construction programme

Check that S/c / Supplier conformance
information has been or is being obtained

Obtain or develop subcontractor Quality
Plans in conjunction with appointed S/c

Check that Document Control has been properly established and everyone is aware of its
method of operation

Instigate system of information request
and appraisal

Establish procedures for inspection
and tests

(13) (15)

Boxes:

Establish
Docu-
ment
Control
(16)

Set up
cost
reporting
system
(14)

Notes:

**See Delivery, Handling and
Storage Procedures**

**See QA/Q for standard
notification and checklist**

]
] **See Quality Plan**
] **Procedures**
]
]
]
]

**See Document Control
Procedures**

See Forms QA/U or QA/V

See QA/W

CONSTRUCTION STAGE

Notes for Page C/4

❑ QA/T Monthly contract report

The report is designed, when added to others, to provide head office management with an overall picture of the company's operations. The report is to be completed in accordance with the calender issued periodically by the area manager. The sections of the report comprise Progress, Resources, Information and finally, and most importantly, a section for the site manager's comments. The responsibility for completing the report on time rests with the site manager. The report forms the basis of the Internal Report meeting with the area manager.

❑ QA/R Agenda for site construction co-ordination meetings

The purpose of these meetings is to establish a regular point of contact with the client representatives. Whilst it is probable that the architect will have his own agenda it is important that the items listed regarding progress, information, inspection and delay are incorporated. The form is to be completed and minutes of the meeting attached when received/prepared.

❑ QA/R(s) Agenda for site subcontractors' co-ordination meetings

The purpose of the meetings is to establish a regular point of contact. They also provide a regular means of checking and recording progress and standards of work and help to ensure a proper flow of information between the parties. Non-attendance by subcontractors should be recorded.

❑ QA/Y Material/Plant Confirmation Receipt

The form is to be used for both plant and materials although each should be listed on separate sheets. Each sheet should be referenced sequentially with the plant sheets prefixed by P and the materials sheets by M (i.e. P1, P2, M1, M2 etc.).

The sheet provides information on the receipt of materials/plant for the accounts department together with a record of its checking and comments on its condition.

❑ QA/Z Estimator feedback

In order to improve the quality of tenders the Estimator requires feedback on the performance of subcontractors, suppliers and consultants that your contract is using. In order to achieve this you are asked to enter a star rating, one to four, each month. In addition if any Bills of Quantities rates are either too low or too high (remember we must be competitive) the Estimators need to know.

Who	Flow	Notes

At regular intervals during construction

Site
Manager
(13)

Site QS
(14)

Construction
Staff
(15)

Maintain records of progress vis a vis
programme

Provide required reports to
Head Office

Regularly update and record
information supply situation

Schedule and carry out
equipment tests

Monitor resource value and
use to date

Monitor deliveries of plant and
materials and inform Accounts

Evaluate performance of
S/cs, Suppliers and Consultants

Monitor and update Quality Plan
and ensure inspection and test
procedures are being complied
with and records are being
properly compiled

Hold Site (Co-ordination) meetings
with Architect / Consultants

Hold Site (Co-ordination) meetings
with Subcontractors including
developing balance of
S/c Quality Plans

Attend regular Internal Report
meeting

(13)

Notes:

]
]See QA/T
]

See QA/T for checklist

See Quality Plan
Procedures

See QA/T

See QA/Y

See QA/Z

See Quality Plan
Procedures

See QA/R for checklist
agenda and notes

See QA/R(s) for checklist
agenda and notes

Site Manager to discuss
contents of the Contract
Monthly Report with Area
Manager

Notes for Page C/6

❑ QA/Z Estimator feedback

In order to improve the quality of tenders the Estimator requires feedback on the performance of subcontractors, suppliers and consultants that your contract is using. In order to achieve this you are asked to enter a star rating, one to four, each month. In addition if any Bills of Quantities rates are either too low or too high (remember we must be competitive) the Estimators need to know.

Who	Flow	Notes

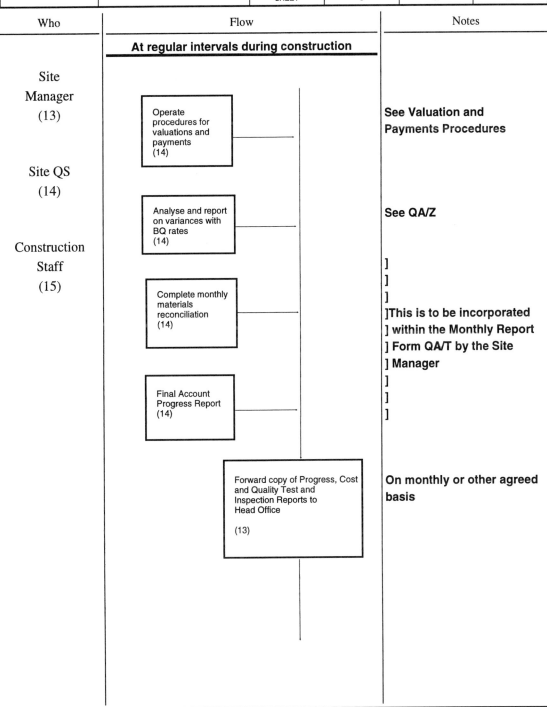

At regular intervals during construction

Site Manager (13)

Operate procedures for valuations and payments (14)

See Valuation and Payments Procedures

Site QS (14)

Analyse and report on variances with BQ rates (14)

See QA/Z

Construction Staff (15)

Complete monthly materials reconciliation (14)

]
]
]
]**This is to be incorporated**
] **within the Monthly Report**
] **Form QA/T by the Site**
] **Manager**
]
]
]

Final Account Progress Report (14)

Forward copy of Progress, Cost and Quality Test and Inspection Reports to Head Office

(13)

On monthly or other agreed basis

CONSTRUCTION STAGE

Notes for Page C/8

❑ QA/Q Site (co-ordination) meeting timetable

The form is designed to provide the basic information required by all parties. Copies of the form providing dates for proposed meetings should be sent to participating consultants/subcontractors as soon as possible. It is recognised that subcontractors will normally attend a separate meeting. An agenda can be found on forms QA/R and QA/R(s).

❑ QA/X Confirmation of receipt of verbal instructions

Most building contracts provide for the formalisation of verbal instructions by the architect by the issue of a written confirmation within a set period. The form allows designated staff (see job descriptions) to confirm such instructions promptly.

Note should also be made of any correspondence which may affect the validity of the written confirmation.

Who	Flow	Notes

On event

Who	Flow	Notes
Site Manager (13)		**See Enquiry Procedures**
	Place balance of subcontract/supply orders	
		See QA/Q for standard notification and checklist
Site QS (14)	Provide new S/c with timetable for meetings	
		See Ordering Procedures and S/c Quality Plans
	Instruct and control subcontractors	
Construction Staff (15)	Update Quality Plans upon the receipt of significant instructions from architect	**See Quality Plan**
	Confirm verbal instructions from Architect	**See QA/X**
	Obtain necessary extension of time/completion certificates	
	(13)	
	Ensure that quality of records produced by others is adequate for purposes of payment / recovery (14)	
	Maintain necessary QS records for payment / recovery (14)	

CONSTRUCTION STAGE

Notes for Page C/10

There are no notes for this page

Who	Flow	Notes

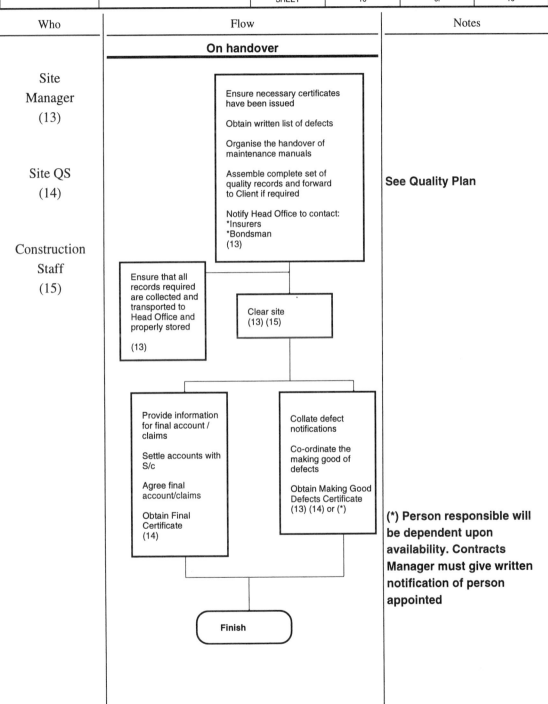

On handover

Site Manager (13)

Site QS (14)

Construction Staff (15)

Ensure necessary certificates have been issued

Obtain written list of defects

Organise the handover of maintenance manuals

Assemble complete set of quality records and forward to Client if required

Notify Head Office to contact:
*Insurers
*Bondsman
(13)

Ensure that all records required are collected and transported to Head Office and properly stored

(13)

Clear site
(13) (15)

Provide information for final account / claims

Settle accounts with S/c

Agree final account/claims

Obtain Final Certificate
(14)

Collate defect notifications

Co-ordinate the making good of defects

Obtain Making Good Defects Certificate
(13) (14) or (*)

Finish

See Quality Plan

(*) Person responsible will be dependent upon availability. Contracts Manager must give written notification of person appointed

This page left intentionally blank

THE SYSTEM DEVELOPED - GENERAL PROCEDURES

8.0   Introduction

As was noted briefly in Chapter 4 and referred to in the Introduction and Flow Charts of Chapters 5 to 7 we have, for reasons of brevity, continuity or flexibility produced certain areas of the system in the form of General Procedures. The most important examples of these were the Document Control and Quality Plan Procedures. Our discussion of the core of the QA system therefore would not be complete without the inclusion and explanation of these examples.

8.1   Document Control procedures

The principle is to ensure the traceability and control of all documents and their enclosures. In order to achieve this all documents are recorded upon receipt or dispatch and given a reference. In addition the circulation of the documents is also recorded so the company knows who had what and when. It was noted in the investigation that documents were both received and sent by potentially different groups of individuals at the various stages of the project so there was a need to establish continuity. It was also recognised that establishing the full document control system for all tenders was potentially wasteful of resources. Therefore at the tender stage a modified version of the procedures shown hereafter was produced, establishing continuity in its use of codes but allowing less complexity in its method of filing.

A particular problem noted in developing these procedures was to design a drawing register which allowed room for recording circulation whilst accommodating the issue of revisions and retaining a sequential listing of drawings. Although the forms illustrated are actually A3 size there will still be problems in establishing how much space to leave between items for a potentially large number of revisions. A better system, if facilities exist, would be to use computer based spread sheet versions of the forms with paper copies retained in a back-up file. The use of such a system will allow the insertion of extra lines between drawing references. The person with responsibility for ensuring that superseded drawings are not used must be established. The control of other documents is established by the site manager within the Quality Plans.

8.2   Quality Plan procedures

The procedures allow the completion of the quality system in a project specific format. Responsibility for its completion has preferably to remain with those responsible for its implementation, hopefully generating thereby a sense of property. The requirements therefore have to be clearly explained but at the same time allow the site manager the maximum flexibility possible. It was recognised that initially there would probably be a need to provide guidance or even assistance in preparing the Quality Plan but the site manager must always remain responsible for the final product. The sequence of general and detailed Quality Plans recognises the reality in terms of time, resources and information flow in most project situations. A good Quality Plan needs to use the best available information and this will not always be available early in a project. This also accommodates the fact that the Quality Plan will be subject to change if significant variations occur.

The Quality Plan procedures are a key element within the Quality System. It is therefore essential that all those responsible for their preparation attend a course of instruction concerning their preparation and receive some explanation and practice in the use of the procedures described later in this chapter.

❏ **A detailed edited extract of this section of the Manual follows.**

All pages with the QA Manual heading are taken directly from the Manual.

Authors' note

The terminology used within the system is the terminology which is familiar to the industry and not necessarily that used within ISO 9000 e.g. both subcontractors and suppliers would be referred to as subcontractors in ISO 9000 but are referred to by their industry title in the extracts of the system.

Document Control Procedures

Generally

The object of these procedures is to increase the traceability of every significant piece of information issued during a project.

All incoming and outgoing documents should be recorded **including their enclosures.**

Mobilisation/Construction Stage

At the commencement of each project, files should be established with the following sections:

A. Head office

1. Records Section.
2. Incoming Correspondence.
3. Outgoing Correspondence.
4. Internal Correspondence.

A master copy of all documents should be kept in these files and not be removed. These should either be originals or you should ensure that photocopies contain all relevant information (e.g. small print on the back of quotations).

B. Site

1. Correspondence to and from architect in separate sections. Minutes of meetings to be filed in a separate section.
2. Correspondence to and from subcontractors in alphabetical order. Instructions to subcontractors filed in separate sections.
3. Correspondence to and from materials suppliers in alphabetical order.
4. Correspondence to and from plant suppliers in alphabetical order.
5. Architect's Instructions, Clerk of Works' directions and Confirmation of Verbal Instructions in separate sections.
6. Requests for information and replies.
7. Quotations from possible future subcontractors and suppliers or plant hirers.
8. Records (a copy of all records to be retained on site).
9. Tender make up and costing information.
10. Applications, valuations and certificates.
11. Internal correspondence and safety.

12. Contract Documents File (a single copy of **ALL** contract documents should be kept in this file). Note - the contents are only to be listed in the Quality Plan file.
13. Quality Plan file, refer to the Quality Plan procedures for contents.
14. Master Index file.

The face of the file/files should be marked with the Contract Number, Project Name and contents in accordance with the above. Continuation files should be numbered sequentially dated from * and to *. Completed files should be marked as continued on File No. * which is the next file.

❑ Form DC1 Correspondence Index

This form is used first to create the Master Index of all correspondence received or sent. It should also be used as a standard index at the front of each of the files or sections noted previously. The brief information it contains can be used to trace particular items of correspondence either from the Master Index or the Section Index. Note the form is designed for use in recording either receipts or dispatches.

DATE	CODE	FROM/TO	ITEM REF	ITEM DATE	DESCRIPTION	ENCLOSURES	CIRCULATION
28/03/9*	S/MC/29	B.G. Supplies	Bg / 11	24/03/9*	Hardcore quote	None	Agent/file
01/04/9*	A/MC/26	A.N. Arch	An / 2	28/03/9*	Instruction	See DC4	Agent/QS/file
01/04/9*	Sc/MC/44	Robins & Dark	Rd/ 10	27/03/9*	Steelwork info	See DC4	Agent/Arch/file

The majority of columns to be completed are self explanatory. The code column is incorporated to enable easy reference as to the nature of the item of correspondence described.

Examples of codes to be used are as follows:

    MC/A   . . Letter to architect/engineer
    A/MC   . . Letter from architect/engineer
    MC/Sc . . Letter to sub-contractor
    Sc/MC . . Letter from sub-contractor
    MC/S   . . Letter to supplier (inc plant hirers)

S/MC . . Letter from supplier (inc plant hirers)
M . . . . Minutes of meeting with architect/engineer
AI . . . . Architect's Instruction
SI . . . . Clerk of Works' Direction/Site Instruction
Q . . . . . Quotation from possible future S/c/supplier/plant hirer
I . . . . Invoice from supplier
S . . . . . Statement of account
RRI . . Reply to request for information
IM . . . . Internal memo
RI . . . . Request for information
CRI . . Confirmation of receipt of verbal instructions
ISC . . Instruction to subcontractor
O . . . . . Order for materials/plant
T . . . . Termination of hire note
SF . . . . Standard forms followed by their reference letter e.g. SFQA/A etc.

An explanation of any non-standard codes should be attached to the Master Index.

An example of the code therefore could be:

A/MC/26

- **A/MC** means an architect's letter sent to the Main Contractor
- **/26** indicates that it is letter number 26 of this code

Where form DC1 requires the Item Ref and Item Date above, it requires the reference and date shown on the letter referred to, whereas the first date column is the date the letter was received by the company.

The records file should contain on conclusion of the construction stage all completed forms noted during the appropriate flow chart procedures plus all tests and inspections detailed in the Quality Plan. A copy of all documents issued should be kept in the Contract Documents file.

If correspondence has a number of enclosures also use Form DC4.

❑ Form DC4 Enclosures Record

Enclosures should always be fully documented. Whenever it is necessary to document enclosures this form should be used and referenced back to the main information shown on DC1.

DATE	CODE	FROM / TO	ENCLOSURES
01 / 04 / 9*	A/MC/26	A.N. Architecture	Drawings 2056/101A, 107C, 112D, 115 F and
			2055/45C
01 / 04 / 9*	Sc/MC/44	Robins and Dark Ltd	Fabrication drwgs 01/1 - 01/10 and Spec pages 1 - 8

Give the full reference of the enclosures in the above **Enclosures** column and use more than one line if necessary. If we look at example A/MC/26 you see on DC1 that this is an instruction with enclosures. For the enclosures on DC4 you would then look for the appropriate date and then the correct code.

Drawing Register

A Drawing Register must be established using Forms DC2 and DC3.

❑ DC2 Master Register

Enables you to establish standard circulation lists for drawings. All individuals or companies to be issued with information are given a number on the Master Circulation List and the number of copies they receive is entered in the right hand column. This portion of the form is completed independently from the Drawing Schedule. When the Drawing Schedule itself is filled in the form is used as a master index for the various sources or series of drawings. These drawings from different sources, series or titles can each be given up to three standard circulation lists as example below:

SOURCE	SERIES NUMBER	TITLE	STANDARD CIRCULATION													KEY MASTER CIRCULATION LIST		No Cop
			A				B				C							
A. Arch	100	Plans	1	2	3	4	1	2	4	5	3	4	5	6		1	Site	5
			8	9	10	15					10	11	12			2	Evans Flooring	1
																3	Jones Tiles	1
A.Arch	100	Elevations	1	5	6	7										4	Whyte Plumbing	1
																5	Smith Windows	1
																6	Acme Cladding	1
																7	Tog Painting	1

In the example the architect's plans and elevations are treated as different titles which will have different circulation lists. In addition each drawing received should be entered on Form DC3.

❑ DC3 Detailed Register

Drawings from different sources or a different series should be entered on separate sheets and each sheet should be numbered consecutively to form a complete register. The Drawing number, title, revision letter and the date of issue should be entered. Circulation can either be a standard circulation list (see DC2) or numbers drawn from the Master Index on DC2. With each revision this circulation can be amended.

DRAWING NUMBER	TITLE	REVISION PREFIX / DATE AND CIRCULATION					
		REV DATE	CIRC.	REV DATE	CIRC	REV DATE	CIRC.
100 / 10	Ground floor plan	A 08-06-9*	A	B 23-07-9*	B		
100 / 11	First floor plan	A 08-06-9*	A	B 30-07-9*	B	C 14-08-9*	1, 2, 3, 6

On the conclusion of the contract the above forms will provide a record of all the drawings received and to whom each and every issue has been circulated.

Note - the person entering revisions on the Drawing Register is responsible for:
1. **Ensuring that the retained copies of the preceding issue are marked superseded**
2. **Noting on the superseded drawing the date they were superseded**
3. **Notifying all those circulated with copies of the previous drawing of the new revision and the date of revision.**

Quality Plan Procedures

Generally

The Quality Plan is drawn up to help the site manager firstly define and subsequently control the quality requirements of the project. Its production requires analysis of the work and the abstraction of imposed quality requirements from the project documentation. In addition the project management team must establish and define their own (and the company's) requirements for the inspection and control of quality.

The Quality Plan is formulated in two stages:

- the general overall plan called the General Plan and
- Detailed Plans which follow the broad outline on the General Plan.

The General Plan shall be prepared prior to the commencement of building operations. On completion of the General Plan a copy shall be forwarded to the Quality Assurance Manager.

The site manager is responsible for the Quality Plan though its preparation may be delegated. The site manager or member of the site staff shall be **named** as responsible for the implementation and monitoring of the Quality Plan.

If you examine the first flow chart, GP/QP2, this indicates the initial processes involved in setting up the Quality Plan. The why, the what, the when and where, the who and the how are explained as follows:

The why may be our own aims, statutory, local authority or client requirements.

The what requires that questions are asked. For example is it a key element, does the work require close control, does the specification require inspections or tests etc?

The when and where will be identifiable from the programmes and thus these must be prepared prior to, or in conjunction with, the Quality Plan.

Who does it The site manager is always responsible for the Quality Plan but may delegate another to produce it.
Note the input that may be required from subcontractors.

How it is to be tested or inspected may be dictated by a number of sources; examples of these are listed on the flow chart.

All of the above are brought together in the Quality Plan.

THE REASONS FOR SETTING UP THE QUALITY PLAN

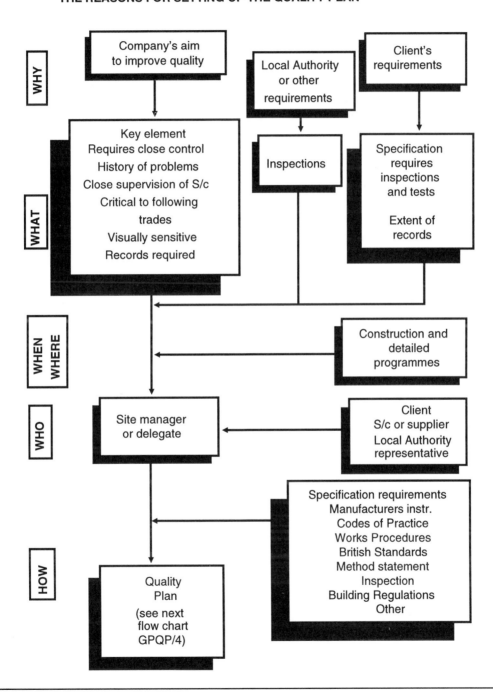

WHY

Company's aim to improve quality

Local Authority or other requirements

Client's requirements

WHAT

Key element
Requires close control
History of problems
Close supervision of S/c
Critical to following trades
Visually sensitive
Records required

Inspections

Specification requires inspections and tests

Extent of records

WHEN WHERE

Construction and detailed programmes

WHO

Site manager or delegate

Client
S/c or supplier
Local Authority representative

HOW

Quality Plan
(see next flow chart GPQP/4)

Specification requirements
Manufacturers instr.
Codes of Practice
Works Procedures
British Standards
Method statement
Inspection
Building Regulations
Other

The Quality Plan

The Quality Plan is first prepared in general terms. Detailed Plans are then prepared, in sections, with the first priority given to the early operations or operations with a long lead-in. The detailed sub-programmes for these early operations must be prepared to enable a clear examination of the precise work involved.

Refer now to the flowchart GPQP/4 on the next page.

As you will note the first task is to draft a complete schedule of the drawings, specifications and other documents from which the plan is prepared.

This schedule of documents, together with your own requirements and any subcontractors' plans which may be available, form the data source from which the plan is prepared. The contract programme then tells you when and where and enables the completion of the General Quality Plan using form QP1. In addition any necessary or available sections of the Detailed Quality Plan can be completed using form QP2 supported by the Test and Inspection Plan Form QP3.

Having prepared the Quality Plan, items such as hold points should then be marked on the construction programme. This acts as a visual aid in monitoring their compliance relative to progress. This process should, if possible, be completed during the mobilisation/pre-construction period. However it is accepted that it may also span the construction period itself. In addition it is almost certain that further Detailed Quality Plans or amended Quality Plans will be prepared at intervals during construction.

The forms, QP1-3, are designed to relate to the operations (by name, number or both) shown on the construction programme or sub-operations shown on detailed sub-programmes.

Look first at the General Quality Plan on the next but one page reference GPQP/5.

SETTING UP THE QUALITY PLAN

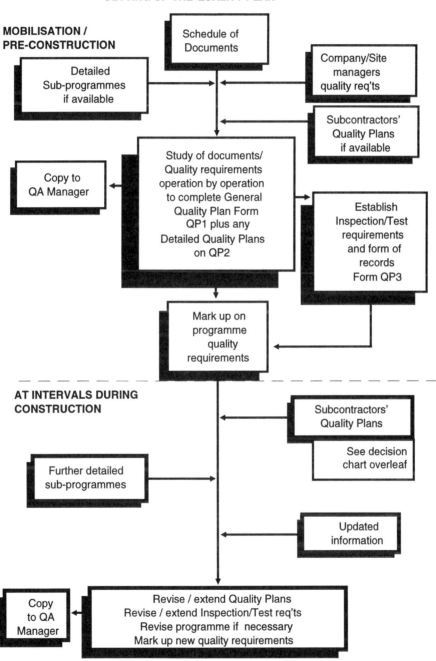

❑ QP/1 General Quality Plan

QA MANUAL	General Quality Plan	SECTION	Company Procedures		
		STAGE	General		
		REF	QAM 4.2.4	QP1	
		REV		DATE	
		SHEET		of	

Project name			Ref		Date	
Operation/ sub-op	Detailed QP ref	Spec. reqm'ts	Client/3rd party approval	MC control type	S/c QP reqd	Other
10 Substructure						
13 Hardcore	Sheet 4	BQ Preambles	Yes	Certificate	-	-

Operation/sub-op - allows you to list the operations as known at that stage in sequence allowing space for additional operations if they are thought likely.

Detailed QP ref - makes provision to cross reference to the Detailed Quality Plan when prepared.

Spec. reqm'ts - allows you to note in general terms the source of the control requirements.

Client/3rd party approval - allows a simple yes or no in terms of whether the control requirements also involve client or another third party's approval.

MC control type - i.e. a test, an inspection, a certificate, a method statement etc.

S/c QP reqd. - recognises the input which may be necessary from subcontractors in certain operations or sub-operations.

Other - allows space for any other points you may consider relevant.

Having worked through all the known operations and utilised the general information available it is then necessary for you to consider the availability of information for sections of the Detailed Quality Plan.

❑ QP/2 Detailed Quality Plans

These are generally formulated from detailed sub-programmes but may obviously be taken from the construction programme if it is sufficiently detailed. Individual sections of the Detailed Quality Plan **must** be completed before the work they are related to is constructed, and immediately forwarded, together with any amendments to the General Quality Plan, to the Quality Manager.

QA MANUAL	Detailed Quality Plan	SECTION	Company Procedures		
		STAGE	General		
		REF	QAM 4.2.4	QP2	
		REV		DATE	
		SHEET		of	

Project name				Ref			Date	
Operation/ sub-op	Spec reqm'ts		MC reqm'ts	S/c/ Sup/MC	Test /Ins	Hold point	Client/3rd party	Other
	Doc ref	Code	code	QP ref	ref	ref	approval	reqm'ts
10 Substructures								
13 Hardcore	BQ 2/56C	C2	-	-	Sheet 2	13/1	Yes	-

Form QP2, the Detailed Quality Plan, should follow the general plan in terms of format but may obviously expand the number of operations or sub-operations. Once sub-operations have been added on the Detailed Quality Plan they should also be referenced within an amended section of the General Quality Plan. This second form allows more room for detail.

Operation/sub-op - has the same function as on the Form QP/1 - to identify by name or number the operation.

Spec reqm'ts - this column requires **precise** information rather than the general information on QP1: firstly of the document reference from which the specification requirement comes; secondly to identify the nature of that requirement in the form of a code e.g. T1 - test type 1, C1 - certificate type 1 etc.

Note - a key to these codes must be included with the plan.

MC reqm'ts - indicates that the imposed quality requirement is our own and should be entered using a similar system of codes. These will include references to method statements and works procedures which can be developed from the guidance notes for works procedures. Examples attached are Sheets SD/WP/30 and SD/WP/14.

SD/WG/30 refers to drainage runs listing typical activities to be considered and the general points at which control by the use of a hold point could be established. These are of course general lists of both activities and hold points and must be applied repeatedly to the individual lengths of drain runs present to build up a total picture of this section of the Detailed Quality Plan.

Similarly other works procedures such as SD/WP/14 Brick/Blockwork (General) lists activities and hold points and may, in their case, be divided into grid references around and within the building.

S/c/Sup/MC QP ref - a significant number of operations are likely to be sublet so it may be necessary to prepare with, or obtain from, a subcontractor or supplier a section of the Quality Plan. This is discussed in more detail later but suffice it to say at this point that you cross reference your Quality Plan here to any subcontractor or supplier Quality Plan.

Test/Ins ref - allows you to enter a cross reference to the Test and Inspection Plan which expands the detail of this section on Form QP3 which will be considered shortly.

Hold point ref - allow you to establish inspection points within the construction. They operate using Form QA/Y in preventing progress without approval. They are designed for use by subcontractors and our own labour. Hold points are identified by the operation number and sequentially i.e. 23/10 would be the 10th hold point for operation No. 23.

Client/3rd party approval - allows you to note any client or third party approval normally as a Yes or No against the more detailed list of preceding items.

Other reqm'ts - again allows space for points considered relevant.

The final form for use in this process is the Test and Inspection Plan QP/3.

❑ QP/3 Test/Inspection Plan

QA MANUAL	Test / Inspection Plan	SECTION	Company Procedures		
		STAGE	General		
		REF	QAM 4.2.4	QP3	
		REV		DATE	
		SHEET	of		

Operation	Test / inspect		Staff responsible	Records		
sub-op	Type	When		Type	To	Stored
10 Substructures						
13 Hardcore	Sulphate content	Before	Site manager	Certificate	Client/Site	Site/head office
		commencement			head office	

Operation/sub-op - again the form starts with the operation name or number.

Test/inspect - gives details of the test or inspection required:

- the description of the type of test
- when and where it will occur i.e. daily, at the end of the operation etc.

Staff responsible - names the person responsible for:

- ensuring that the requirement is properly executed
- providing the appropriate records.

Note - this person could be external to the organisation if no other option is possible.

Records - establishes:

- the type of records e.g. completed form, certificate, marked up drawing etc.
- where the records go e.g. copy in site file, copy to head office
- where they are stored.

Subcontractors and suppliers

As was noted in completing the forms a substantial proportion of work is likely to be sublet so it may be necessary to cover these sublet elements with the subcontractors' own Quality Plans. The final diagram GPQP/10 is a decision chart which assists you in deciding the route to be taken in obtaining a Quality Plan from a subcontractor.

Looking at the diagram GPQP/10 the first decision is whether your subcontractor or supplier actually needs their own separate plan.

If **NO** you inspect and approve their work yourself within your own test and inspection plan. If **YES** you must first discover whether the subcontractor or supplier operates their own QA System in accordance with ISO 9000. This information is available on Record Form QA/F. If they do have their own QA System you request them to submit their Quality Plan for approval. If **NO** you should enquire with the subcontractor whether they require to use our standard planning forms QP/2 and 3. If **NO**, again they prepare it in their own format in accordance with the principles set out in Section 7 of the Quality Plan file noted on GPQP/14 hereafter.

Note - the points listed in that section numbered 1 to 10 form part of the standard enquiry conditions and are also incorporated in their subcontract conditions. If it is **YES** to using our standard forms copies are provided with a note that column 5 is to be left blank; we will then use this to cross reference to our own Quality Plans.

It must be recognised, at this stage, that in addition to setting out the imposed quality requirements within their subcontract, we must also be prepared to offer some guidance to certain subcontractors in their preparation of their Quality Plans.

The final stages are to check that the particular requirements notified within the enquiry and subcontract documents and listed in Section 7 hereafter are incorporated in their Quality Plan; finally to inspect and to approve the subcontractors/suppliers work at the hold points established with them and noted within their agreed Quality Plans.

DECISION CHART FOR PREPARATION OF DETAILED QUALITY PLANS FOR SUBCONTRACTORS/SUPPLIERS

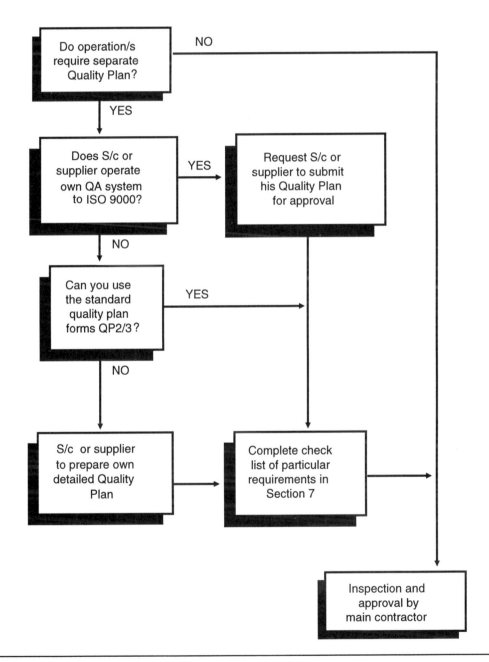

The contents of the Quality Plan file

The Quality Plan for each project shall be held in its own file divided into the following sections:

1. Project particulars (based upon Form QA/A and updated).
2. Schedule of contract documents - checked monthly or at shorter intervals to be specified by the site manager.
3. Site organisational structure and procedures - initial and updated.
4. Contract programme and detailed sub-programmes.
5. General Quality Plan - initial and updated.
6. Detailed Quality Plans.
7. Subcontractors' detailed Quality Plans.

Amplification of Quality Plan section contents

1. Project particulars

This section is based on checking and updating Form QA/A and should include the names and addresses of the principal parties to the contract and the design team. It should in particular state their representatives having responsibility for quality.

2. Schedule of contract documents

This will **list** the documents comprising the contract, typically the Employer's bills of quantities, specification(s), drawings and any related correspondence.

3. Site organisational structure and procedures

This is to comprise the following.

3.1 Charts

1. Area management structure chart showing lines of responsibility from site manager up to area manager including any off-site services.
2. The client and design teams' organisational structure as related to the project in as much detail as possible.
3. A site management structure chart showing job functions and lines of responsibility. This should include all staff having a supervisory or administrative role.

3.2 Schedules

1. Showing names, addresses and telephone numbers for each person shown on the charts.
2. Giving job descriptions. The duties of all administrative or supervisory staff where at variance with the standard job descriptions. See also Form QA/P and update.

3. Showing subcontractors and suppliers names, addresses, telephone numbers and personal contacts.
4. Showing the programme of formal meetings with the client / subcontractors as Form QA/Q and updates.

3.3 Written procedures
1. For communicating instructions, variations and information to the point of work.
2. For distribution and registration of drawings (also see Document Control Procedures).
3. For recording of telephone conversations, verbal instructions (see QA/X), informal meetings.
4. For obtaining third party approvals.

4. Contract programme and detailed sub-programmes

This section incorporates a copy of the latest and all previous contract programmes clearly marked as superseded. Also included are procedures for the monitoring and review of the current programme.

Procedures must be included for the preparation of detailed and sectional programmes, weekly work plans etc. Information must be included to indicate dates for receipt of design information and placing of orders, allowing a suitable time period for inspection of information.

Note - Form QA/V should be used taking the document control reference shown on DC1 and 2 and allowing a record to be kept of its appraisal.

5. General Quality Plans

Each operation on the construction programme should be examined and a plan drawn up for its effective control.

The complexity of the plan will be dependent upon the amount of information available, at this early stage, the type of work and whether it will remain available for inspection. Other factors to consider for this initial plan are:

- does the specification call for inspections and tests?
- is it a key element? does it require close quality control?
- is there a history of past failures or problems?
- how critical is it to following trades?
- will it be difficult to gain access to later?

GENERAL PROCEDURES

- does the subcontractor carrying out the work require careful supervision?

6. Detailed Quality Plans

As the contract progresses, more information becomes available or variations are issued and the general plan can be expanded and updated. Detailed Quality Plans must be prepared prior to the commencement of each operation and a copy of both detailed and revised General Plan must immediately be forwarded to the QA Manager.

A Detailed Quality Plan will consider:

1. Preparation and review of detailed programme.
2. Specific requirements for work guidance notes.
3. Specific requirements for design and materials approvals.
4. Records to be completed and returned to the main contractor.
5. Use of check sheets and requests for inspection (see Form QA/W).
6. Inspection and test requirements.
7. Can we use statistical techniques to establish test/inspection requirements. Identification of the subcontractors' person-in-charge of site operations.
8. Any samples or mock-ups required.
9. Requirements for commissioning and provision of as-built drawings and/or maintenance manuals.
10. Quality Plans required to be prepared by subcontractors.

It is also necessary to accommodate work by subcontractors by incorporating their Quality Plans.

7. Subcontractor Quality Plan

This plan applies to those subcontractors or suppliers who do not have an ISO 9000 Quality System approved by a certification body. Those that do will submit for approval their own Quality Plan.

Subcontractors who do not have their own Quality System should submit a Quality Plan in accordance with the following points (as required by their subcontract agreement):

1. A programme of subcontract works shall be drawn up by the subcontractor and agreed with the main contractor. It shall show the dates for receipt and approval of the supplier's drawings, information, etc. where required under the subcontract.
2. A comprehensive schedule of inspections and tests shall be prepared by the subcontractor. Acceptance standards shall be ascertained by reference to the specification.
3. The above must be approved in writing by the site manager before work commences.
4. The subcontractor must name the person based upon the site who will be responsible for quality.
5. The points of inspection of work by the main contractor must be noted on the Quality Plan. The work where covered should be scheduled for inspection and approved before it is concealed.
6. Records shall be maintained by the subcontractor and submitted to the main contractor detailing:

    *   conformance of materials to specification
    *   inspection and test certificates
    *   as-built documentation
    *   operating and maintenance manuals.

7. The subcontractor is to use the contractor's work procedures when requested by the site manager.
8. All work must conform with the specification for work to proceed beyond a specified hold point. Only non-conformances covered by a waiver will be accepted into the works. Such acceptance or waivers will be confirmed in writing by the main contractor's specified representative(s) using Form QA/W.
9. Any approvals given on Form QA/W are made with the objective of progressing the work. However, the subcontractor should note that all work is subject to final inspection and client approval at date of handover of the works.

QA MANUAL	Drainage runs	SECTION		Supporting Documentation	
		STAGE		Mobilisation / Construction	
		REF	QAM 5.2	SD/WP/30	
		REV		DATE	
		SHEET	1	of	1

Inspection/ Hold point ⟍ Activity	Pre-commencement	Before starting excavation	Before laying bedding	Before laying pipes and regularly during laying	Before placing concrete, surround or backfill	Periodically during backfill operation	Before completion	Signature / date
Materials : Compliance with spec. Certificate	●							....
Setting out control: Line/level		●						....
Permit to dig: Service clearance obtained		●						....
Information to gang: O/s to C.L. depth to invert and exc. Pipe type, size. Branches type, depth bedding surround		●						....
Excavation: Suitability, width, depth of trench, formation, support			●					....
Bedding: Depth, material				●				....
Pipe laying, jointing: Depth of bed, line, level, invert supported, correct type, dia. of pipe, number and size of branches				●	●			....
Initial test: Type					●			....
Surround: Compaction, pipe, line/level OK? Branches closed off and location recorded						●		....
Backfilling: Compaction? Material?						●		....
Final test: Type, flushing and proving of lines and branches							●	....

Activity \ Inspection/Hold point	Pre-commencement	Test panel	Post commencement - daily	Up to DPC	Every 1.2m vertically	At cill, head, jambs of all openings	Before closing all cavities	At installation of cloaks, trays, flashings	Before striking scaffolding	Signature / date
Materials: Compliance with spec. Certificate	●									....
Setting out: Dimension check Line/level/plumb Position of openings, features etc.	●			●		●				....
Building tolerances: Bricks, overall, openings	●									....
Appearance: Brick, mortar, joint width, pointing *(Cleanliness, uniform colour, texture etc).*		●	●		●				●	....
Mortar mix: Batch controls?			●							....
Beds and Perps: Filled?			●	●	●		●			....
Ties: Number/spacing, type, cleanliness, slope/drip, position *(General/corners/openings)*			●		●	●	●			....
Insulation: Cut/fixed correctly Cavity gap			●		●		●			....
Cavity: Width, clean, boards/battens in use, sand beds for cleaning			●	●	●		●			....
Dpc/trays/flashings: Position, fixing, continuity joints, correct type/detail				●			●	●		....
Weepholes: Position, number, clear?						●	●		●	....

This page left intentionally blank

MAINTAINING THE SYSTEM

9.0   General

The development of the system should also encompass provision for its maintenance and supervision. The major tool which provides for this purpose is auditing. It would be wrong to view auditing as an inspection which is only necessary because of the need to maintain the Quality System. Auditing is a useful tool for management in maintaining general contact with the conduct of the business. It is also important that the emphasis of maintaining the system remains focused on improving the quality of what the company does rather than simply increasing the rigour of procedures. One of the functions of auditing is to identify procedures which are ineffective i.e. too costly, too time consuming or simply unnecessary.

9.1   Auditing

A detailed guide to Quality Systems Auditing can be found in ISO 10011 : 1991 which specifies objectives, responsibilities, roles and audit procedures. This standard is suitable for both internal and external audits.

A quality audit is defined in IS0 10011 -1 as - *a systematic and independent examination to determine whether quality activities and related results comply with planned arrangements and whether these arrangements are implemented effectively and are suitable to achieve objectives.*

The audit itself must be systematic in the way it is organised and performed. Procedures must therefore be included as part of the system.

9.2   Types of Audit

There are three types of audit which normally occur: third, second and first party audits.

9.2.1 Third party

If the system developed is certified the third party audit will be undertaken by the independent body with which the company's scheme is registered. The particular requirements of their audit will vary between certifying bodies but the principle is to ensure that the company's system remains in compliance with the International Standard.

9.2.2 Second party

These are carried out by the company on their subcontractors or suppliers. Such audits help the company by providing information to assist selection process and also increase mutual understanding and awareness of quality requirements between the company and its suppliers or subcontractors.

9.2.3 First party

First party or internal audits are those that an organisation carries out on itself to confirm to executive management that the system is working effectively. Reasons for such audits include:

- QA standard requires it (see Clause 4.17)
- control mechanism utilised by management (note that IS0 9002 requires the result of internal audits to form an integral part of the input to management review (see Clause 4.1.3)
- to correct non compliance before third party audits find them.

BS 5750 : Part 4 : 1990 *(Authors' note - this is due for harmonisation with ISO 9000 series but remains current at this time)* provides a definition of internal audits, under Clause 4.16, which states that such audits are designed to ensure:

- that the quality system documentation adequately defines the needs of the business
- that the documented procedures are practical, understood and followed
- that the training is adequate.

Other reasons may include:

- that the procedures are effective in improving quality
- that they provide the auditor with an opportunity to improve the quality system
- that they meet regulatory requirements.

Essentially the internal audit checks the system, whether it is comprehensive enough, whether the procedures work (in improving quality) and whether they are being used. The training aspect ensures that those involved in operating the system and producing the work are adequately qualified to do so. Training will be examined in the next chapter.

9.3   What to audit

As was stated in Chapter 2, ISO 9000 requires that the system should define a planned sequence of internal audits (see Fig 9.1 for example of audit programme). It is however recognised that certain elements within the system are more important than others and should be given more frequent attention. The emphasis given within these procedures was that:

1. An examination and check list be completed for all Quality Plans and amended plans within two weeks of their submission by the site manager.
2. A monthly spot check of head office files be undertaken to ensure that the appropriate forms relative to the stage of progress have been adequately completed. This check would cover a third of all current projects, all of which would then be covered over a three month period.
3. An initial visit be made to all projects within their first three months to check the adequacy of the Quality Plan vis à vis:
   - contract documentation
   - progress.

During this visit checks would also be made of the site's Quality Plan file (see Written procedures Chapter 8 Section 3.3 page 121) and Document Control procedures.

Within the system this initial visit is described as a guidance session which results in the preparation with the site manager of a joint report forwarded to the area manager.

Details of these checks together with other less frequent checks and a notice of corrective action would then form the basis of a monthly quality report from the Quality Manager to the area manager. This should be designed to keep executive management's finger on the pulse of the business rather than purely a check on quality.

The system also defines through its audit procedures that new audit programmes are to be agreed and recorded with the area manager every three months. These programmes can also vary the frequency of checks set out in the audit procedures (not detailed in this book).

Audits are carried out for one or more of the following reasons:
   - to evaluate suppliers and subcontractors before entering into a contract with them
   - to verify that the company's own quality system continues to meet specified requirements within a particular contractual relationship
   - to evaluate the company's own quality system against the ISO 9002 Standard.

9.4   When to audit

1S0 9002 states - *Internal quality audits shall be scheduled on the basis of the status and importance of the activity to be audited.*

Status relates to the maturity of the activity and its relationship to other activities in the process, e.g. there is no point in auditing an activity if it has only just commenced or if it plays a very minor part in the overall process.

Importance can be established in terms of the effect of a non-compliance with the planned procedures e.g. incorrect specification of foundation concrete could be extremely costly to remedy if wrong. Audits are necessary mechanisms to confirm that controls are in place to detect such errors in good time.

Using status and importance as criteria a programme can be prepared which should be comprehensive enough to cover the whole system once every year. An example of this may be as shown in Fig 9.1 below.

Departments/projects	Months of the year											
	Jan	Feb	Mar	Apr	May	Jun	Jul	Aug	Sep	Oct	Nov	Dec
Purchasing												
Project 1												
Estimating												
Project 2												
Technical												
Project 3												
Planning												
Project 4												
Accounts												
Project 5												
Surveying												
Project 6												
Suppliers/subcontractors	at regular intervals throughout year											
Project start up visit	as and when new projects commence											

Key	Planned audit date		Agree corrective action	
	Audit complete		Confirm correction implemented	
	Divergences agreed			

Fig 9.1 Audit programme

As with any plan the audit programme needs to be flexible to accommodate change throughout the period (e.g. the audit may reveal significant problems within a project or department which will require further audits which will in turn change the overall audit programme.

9.5   Who should carry out the audit

1S0 9002 states audits - *Shall be carried out by personnel independent of those having direct responsibility for the activity being audited.*

Auditing is one of the functions of the Quality Manager (QM) whose other functions are discussed in the next chapter. In terms of auditing the QM's functions are to:

- ensure the company's audit procedures are implemented
- develop auditing procedures
- ensure audit reports are processed
- maintain audit documents
- plan audit programmes
- summarise audit results for management review

Clearly whilst the QM should carry out audits it is not always practical or desirable for the QM to undertake them all. It will therefore be necessary to select and train staff from within the company who will be responsible for undertaking or assisting in internal audits. The foremost qualities that should be considered when selecting these auditing assistants are:

1. An ability to communicate. Auditors require the ability to produce good reports and also need an understanding of the communication processes within the company.
2. Possess tact and a diplomatic approach. Having to function across departmental and project boundaries means persuading staff to co-operate with audits. To do this they must deal with the problems of internal 'politics' and need to be tactful and diplomatic.
3. Must be independent. If possible it is important that individuals are not involved in auditing their own projects or departments.
4. Not be specialists. Need not specialise in the activity being audited but have an understanding of what is going on to ensure that the system is being adhered to.

When the individuals have been selected the Standard requires that they are trained. BS EN 10011 Part 2 gives guidance on qualifications for system auditors and there is a plentiful supply of commercial courses available for internal auditing techniques.

ISO 9000 requires that results of the audits must be documented and notes the following:

- the deficiencies found

- the deficiencies found
- the corrective and preventive action required
- the time agreed for corrective and preventive action to be carried out
- the person responsible for carrying out the corrective and preventive action.

The audit of the system will probably reveal deficiencies within the system and require corrective and preventive action. The responsibility for these actions with respect to the system, is placed with the QM. Responsibility for corrective and preventive action due to non-use of the system is placed elsewhere (estimator, site manager etc.) but authority to instruct the correction of the deficiency is again placed with the QM. It is recognised that with such authority vested in the QM, its application should be exercised with care to avoid alienating those responsible for using the system.

9.6   The stages of carrying out an audit

Audits should be carried out against the procedures as defined in the quality manual. The audit process can normally be split into four stages:

1. Plan
2. Execute
3. Report
4. Correct

9.6.1 Plan

This would include:

- meetings between QM and auditors to discuss the scope of the proposed audit
- reviewing any previous audit reports for that project, activity or department
- preparing an audit check list. These should relate to the technical detail to be audited. Auditors will also need to note general observations to enable most benefit to be gained from audit process. The check list can also include questions to be asked for staff operating the system as well as recording evidence from the records
- setting dates and times for meetings/visits to project or department to be audited ensuring they are convenient and do not clash with other vital business activities. Executive management should not allow the postponement of audit work except for the most important of reasons as ISO certification depends on maintaining internal auditing.

9.6.2 Execution

The basic role of the auditor/s is to establish what ought to happen, as described in the Quality Manual and Procedures, and then find what actually happens in practice. In general terms this involves:

- observing
- questioning
- seeking documentary evidence

The process should be focused on those responsible for and those undertaking the activities being audited. When problems are discovered the auditor should only note that the discrepancy exists and if possible why it exists by agreement with the personnel being audited. The audit process is not intended to look for answers or to apportion blame. This process comes after the audit is concluded. It is to be expected that discrepancies will be found but not all are to be evidence of 'system breakdown'. They can however provide the means by which improvements to the system can be made.

9.6.3 Report

The report should make clear the auditor's findings and should include:

- date of audit
- name of auditor/s
- scope of audit
- record of discrepancies found
- further observations or comments from staff/auditors.

The report should be signed by the auditor/s and actioned by the Quality Manager.

The results of internal audits must be recorded and such records become part of the quality records required by Clause 4.16.

9.6.4 Corrective action

ISO 9002:1994 states - *The results of the audits shall be recorded and brought to the attention of the personnel having responsibility in the area audited.*

As noted previously the management personnel responsible for the area are the people to take timely corrective action on deficiencies found during the audit.

The stages for corrective action are:

1. Problem identification.

2. Finding and recommending a solution.
3. Deciding whether to implement the recommendation.
4. Either amend system or improve the implementation of procedures.

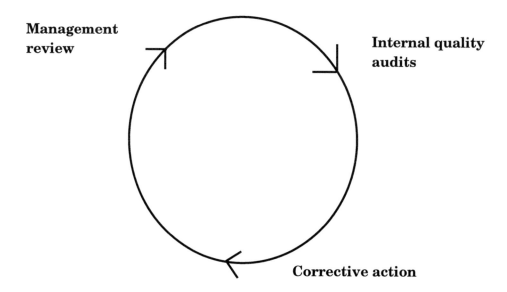

Management review

Internal quality audits

Corrective action

Fig 9.2 Auditing in the continual improvement process

The audit procedures should clearly define management's responsibility for taking corrective action and should also state what is regarded as timely. An important aspect is the necessity of consulting the staff concerned before taking corrective action. This reinforces the principle we established earlier (see Chapter 3) that the system should belong to the staff implementing it. If the audit system is responsive and effective it will be seen by staff as a mechanism for improving what they do rather than a 'checking process' to be feared.

THE SYSTEM DEVELOPED - OTHER ISSUES

10.0   General

Having examined most elements of the system developed we need also to consider some other issues which have been briefly referred to but not detailed within areas of the system discussed. These issues are:

1.   The Quality Manager.
2.   Recruitment, education and training.
3.   Familiarisation.

10.1   The Quality Manager (referred to in the standard as the Management Representative)

10.1.1 Why

The necessity for such an individual stems from ISO 9000 : 1994 Clause 4.1.2.3 that requires executive management to appoint or nominate a management representative with defined authority and responsibility for ensuring that the requirements of the Quality System are implemented and maintained. The planned sequence of internal audits is part of that maintenance function and logically requires that the management appointee or nominee should be responsible for it. Furthermore the audit Clause 4.17 requires that the auditor should be independent of the project structure and report directly to senior management. It therefore appears to be an extension of this logic that the functions of the Quality Manager should be:

- ensuring compliance, maintenance and development of the system
- responsibility for planning and implementing audits
- overseeing the adequacy of the company's training
- familiarising staff with the concept and use of the QA system
- encouraging and facilitating quality improvement.

10.1.2 Who

Bearing in mind these functions what sort of person should a company be looking for.

There may be a tendency to look for someone with a depth of knowledge and expertise in the technical or practical area of construction. A closer examination of the role of the individual seems to indicate that the skills required are primarily communicative and organisational. Within the particular system that we have described the level of practical or technical expertise, whilst present, need not be the primary qualification. Other systems, if they place a greater responsibility for the work produced upon the Quality Manager, create a greater need for technical knowledge. However both ISO 10011 and ISO 9000 indicate the need to separate the management of the system from the management of the physical work to ensure that those responsible for the work remain responsible. Finally it must also be recognised that there is a need for training such an individual. This should be directed not only to their audit function but also to ensure executive management's general responsibility under Clause 4.1. This Clause requires management to ensure each person's ability to undertake whatever role is assigned to them.

It is normal for the person chosen to be drawn from the industry with a background of skills and experience in a particular aspect of the company's work. The person must then undergo a cultural change to disengage themselves from their former role. Their new role is to oversee and advise on the application of quality procedures whilst leaving operational decisions to the staff responsible. Clearly training needs will vary depending upon the individual's background and can include:

1. Quality management courses organised by the Institute of Quality Assurance.
2. Lead assessor/internal auditor courses organised by a variety of certifying bodies and consultants.
3. Courses on communication and personal relations skills.

To introduce the concept of responsibility being rested in the person undertaking the work whilst checking adherence to the system will require sensitivity and skills in personal relations.

This must mean a management style which seeks co-operation and has concern for the people involved. Those companies with an aggressive confrontational style of management will fail to gain support from their staff and will cause problems for the Quality Manager.

Reasons for failure by the Quality Manager include:

- picking the wrong type of individual
- failure to understand the purpose of quality management system
- lack of visible support from executive management
- lack of relevant training
- overly comprehensive systems leading to paperwork overload
- failure to maintain credibility
- allocating quality manager's role to an individual with an already overloaded responsibility.

10.1.3 Where

So where do you place this person in the management structure? Clearly this is a management role and as such should have a well defined and clear status within the company's structure reporting directly to the Chief Executive Officer. This will probably be as near to executive level as possible, a departmental manager or equivalent. In part this conclusion is drawn from the observation that the higher the status of the individual the less that person is likely to require, or indeed use, their authority. The Quality Management function need not be the sole function of the individual appointed, although this will vary dependent upon the size of the organisation. If this function is only part of the person's duties then it is important that the other function/s is perceived as commensurate with the status noted previously.

10.1.4 When

The next question is when should you appoint or nominate a person to this role? As was shown in our discussion of the early stages of the implementation of the system there is a role for a Quality Manager as early as the investigation stage. There is a need within any organisation contemplating the development of a QA system, demonstrated in Fig 1.3 (page 10), for an individual at management level who can combine a level of understanding of QA with the knowledge of the functioning of the organisation. In addition to this point ISO 9000 Clause 4.1.2.3 (a) requires this individual, as one of his/her duties to ensure that the Quality System is established as well as implemented. Our discussion on familiarisation later in this chapter reinforces this need.

10.1.5 What

A typical job description for a QA manager might therefore list the following duties:

- guidance and support in preparation of the QA system
- appraisal and approval to changes in procedures
- monitoring complaints, non-conformances, corrective and preventive action
- regular review of quality documentation
- preparation and control of audit programmes
- responsibility for conducting internal audits
- responsibility for conducting external audits of subcontractors and suppliers
- guidance in preparation of project Quality Plans
- reviewing recruitment and training procedures implementation.

Specific tasks which may not have existed prior to introducing the Quality System would include:

1. Management review meetings.
2. Internal quality audits.
3. Document control monitoring.
4. Training progress reviews.
5. Disseminating quality policies to staff.
6. Monitoring corrective actions taken.

10.2 Recruitment, education and training

Recruitment, education and training must be viewed as the management of an important intangible asset. Companies recognise that in order to manage assets, information is required as to their value and condition and much effort is devoted to this end. Clause 4.1 requires management to ascertain both these things; furthermore under Clause 4.18 they are required to record in writing that they have done so, and to retain such records for inspection. Also, where a deficiency in respect of the individual's experience, qualifications or training, with regard to their present or proposed role, is identified, they are required to rectify the deficiency. Clause 4.18 advises that particular attention should be given to new staff, thus recruitment, education and training become closely linked. Procedures written for this section of the Manual for training should also address recruitment where stricter selection of new staff linked to a long term strategy and career structures could assist in improving the skills base of the company. A particular problem relates to existing staff who may lack formal training or qualifications but this should be catered for by formally excluding staff with 'x' years experience in a particular role from these procedures.

The procedures would then address seven areas:

1. Selection methods, interview techniques.
2. Records of checks on references/qualification.
3. Staff records, qualifications and experience.
4. Training records.
5. Education programmes.
6. Career structure plan.
7. Management responsibility via a delegated structure.

The final point notes the need for the individual's own manager to be given responsibility for monitoring that person's career plan. The managers would then, in turn, be monitored by their immediate superiors.

It is generally accepted that people will not yield their full potential without training. Training however should be planned in a systematic and objective manner. It should not be reactive with individuals recieving training on courses that happen to be offered. Training is expensive and must be directed toward your company's needs. This requires that procedures are incorporated to identify your training needs. This should not only address the issue from the perspective of the individual's needs but in terms of the company's plans:

- development plans for introducing new processes and technologies
- project plans utilising new equipment and techniques
- marketing plans which require entry into new areas
- corporate plans addressing new legislation or changes to the balance of the company's activities.

These may all produce the need for changes in job specification which highlight both training and retraining needs. Simply to identify these needs is not enough; there should also be procedures for implementing training and finally for verifying its effectiveness. Each manager should be responsible for planning and assessing the effectiveness of the training of their own staff. Senior management should then be responsible for coordinating these programmes within the company's overall training plans. Training plans should identify the following:

1. The person responsible for coordinating training.
2. The type of training being undertaken.
3. The provider of the training.
4. Outcomes such as examinations or certificates.
5. The location of the training.
6. The period of training.
7. Who will attend.
8. How the outcomes will be assessed.

10.2.1 Maintaining records of training

The standard requires that you record on a person's file details of courses undertaken including dates, duration, and results. Copies of any certificates gained should be kept on file. This file should be kept both centrally and by the individual's line manager. It is useful to cross reference the training an individual receives with a list of individuals who have undertaken each type of training. This will provide a database to managers of individuals with particular skills or qualifications and facilitate selection of staff for particular tasks.

10.3 Familiarisation

As you will have noted we have suggested that this process is something that must take place from the initial stages of the system's development. The reason for early familiarisation is to foster a sense of involvement in a process of change which will inevitably create some anxiety and even resistance. In addition an early introduction to the principles of the system to be produced is essential to allow the two way flow of information necessary for its development. This is also clearly not the only point at which familiarisation is necessary.

Every effort should be made to produce a system which is clear and easy to use. In attempting to do this there is a problem that you immediately encounter. The sheer volume of paperwork necessary to write something which needs no further explanation often defeats the other objective, that of ease of use. A compromise is to accept that the processes of use and familiarisation with the system could be separated. This can be done by devising introductory seminars or training for groups of staff preferably in cross disciplinary teams. This can be done by a regular programme of seminars initially to cover all personnel but later to specifically targeted groups of staff or areas of operation. An example of this was previously referred to in Chapter 8 under 8.2 Quality Plan procedures as a course of instruction in the preparation of Quality Plans. Seminars also provide an opportunity to exchange views on current procedures.

In addition to the seminars it is necessary to set up a programme of continuing debate on quality or quality system problems which could well be linked to a formal feedback process.

11.0   Introduction

This chapter illustrates edited copies of all forms referred to in the Flow Charts or General Procedures described in the preceding chapters.

As stated the forms are mainly checklists serving to add flesh to the tasks outlined in the Flow Charts or described in the Procedures. Alternatively they provide a list of the points required to be raised within the agendas of meetings that occur at particular stages. The forms when completed will be stored in the Records Section of the filing system described within the Document Control procedures, see Chapter 8.

Our research has shown that the simple act of completing and storing a form known to contain particular information can save considerable time and paperwork. An example of this can be illustrated by reference to the first form, QA/A the Enquiry Report, on the next page. The form sets out the basic tendering information of project particulars, contract details, methods of tendering and staff involved. This information is required at frequent intervals during each of the stages e.g. tender preparation, handing over stage, introduction of subcontractors, completion of subsidiary documentation etc. Checking past project records revealed that the information shown had been abstracted from source and written down on numerous occasions, sometimes by the same individual. This fact is true of many other sets or pieces of information.

Although completion of the forms appears at first sight to be an addition to the paper mountain already generated, this is not so. The forms, rather than being a generator of paperwork, are a known data source which saves abortive time and paperwork spent on tasks which are often needlessly repeated.

Examination of the forms will reveal that they do not involve additional tasks, merely the recording of those tasks which should be completed in the course of the individual satisfactorily executing normal duties.

Authors' note

The terminology used within the system is the terminology which is familiar to the industry and not necessarily that used within ISO 9000 e.g. both subcontractors and suppliers would be referred to as subcontractors in ISO 9000 but are referred to by their industry title in the extracts of the system.

PROJECT PARTICULARS

Project description	Project address

Date enquiry received	Signed	Date tender to be submitted

Project type		Client
		Contact
	Tender ref:	Telephone
Architect		Engineer
Contact		Contact
Telephone		Telephone
Quantity Surveyor		Drawings available at
Contact		By arrangement with
Telephone		

CONTRACT DETAILS

Form of Contract		Method of measurement	
Amendments		Insurances	
Special conditions			
Fixed price	Period	Bond	
Fluctuating-type	Clause no.	Base month	Non-adjustable%
Period certificates	Monthly/other	Period for payment..... days	
Retention	Certified value%	Materials %	Limit of Retention
Liquidated damages	£.....................week/day	Defects liability period	..................months
Commence date	Complete date	Sectional completion	

METHOD OF TENDERING

Traditional	Form of tender	Design and build	Feasibility Report
	Bills of quantities		Form of tender
	Schedule of rates		Drawings
			Bills of Quantities
			Schedule of rates
Category A B C D		Type	Feasibility
Initial Category			Budget
Revised Category			Final tender
Drawings to prepare Yes/No		Bills to prepare	Yes/No

ESTIMATOR'S REPORT

Adequacy of information			
Contract period	Own assessment	Any phasing	Mobilisation

STAFF ALLOCATED

Estimator -name/s	Project co-ord-name/s
Planner-name/s	Site staff-name/s
Others-specify	

QA MANUAL	Programme and Method Review Report	SECTION		Supporting Documentation		
		STAGE		Tender		
		REF	QAM 5.1		QA/B1	
		REV		DATE		
		SHEET	1	of		1

PARTICULARS

Project name	Brief description
	Tender Ref:

TIMESCALE

			Comments
Programme time	Specified		
	Anticipated		
Phasing	Client requirements		
	Contractor requirements		
Major gang sizes			
Major plant outputs			

CONSTRUCTION

List of major work packages	Subcontracts	Response	Onerous conditions	Particular attendance

Material/plant problems	Delivery	Advance orders	Special storage	Other

Labour/staff problems	Availability	Transportation	Other

Logistic problems	Access	Services	Tips	Other

Prepared by:	Date	Checked by:	Date

PARTICULARS

Project name	Brief description
	Tender Ref:

CLIENT CHANGES

Nature of change	Date received

ANALYSIS

High rates/ Low rates	BQ ref	Own Rates			S/c Rates	Comment	Amended	
		Lab	Plant	Mats			Yes	No

Attendance/Prelims

High/low	BQ ref	Comment	Response	Amended	
				Yes	No

Prepared by:	Date	Checked by:	Date

QA MANUAL	Enquiry Records	SECTION	Supporting Documentation	
		STAGE	Tender/Mobilisation/Construction	
		REF	QAM 5.1	QA/C
		REV		DATE
		SHEET	1	of

Project name	Ref	Date

NOTES

The choice of company whilst generally being of the enquirer's own preference should also conform with one of the following criteria:

A. A specified company subject to Note 1 below

B. Companies included within the approved list

C. Other companies subject to checking procedure by Estimator (checking procedures QA/F)

Note 1 - companies should be checked by the Estimator. If they fail to conform the client should be notified.

MATERIALS				SUBCONTRACTORS			
Name of Supplier	Cat	4	Date	Name of Subcontractor	Cat	4	Date

MATERIALS				SUBCONTRACTORS			
Name of Supplier	Cat	4	Date	Name of Subcontractor	Cat	4	Date

Project name	Ref
HEADINGS	REMARKS

CONTRACT DETAILS

Site address	
General description	

EXISTING SITE DETAILS

Buildings		
Services	Telephone	
	Gas	
	Water	
	Electric	
Obstructions and restraints		
Fences or hoardings		
Trees, streams etc.		
Access		
Location in relation to:	Road	
	Rail	
	Public transport	
Conditions of traffic on		
highways adjacent to site		
Parking restrictions/nearest available		

SERVICES (Nearest available)

	Telephone	
	Gas	
	Water	
	Electric	
Drains		
	Foul	
	Surface water	

HEADINGS		REMARKS
GROUND CONDITIONS		
Topography		
	Flat	
	Sloping/Flat	
	Sloping	
Borehole details		
	Location	
	Type of surface soil	
	Stability	
	Visible water table	
Soil storage		
	Location	
	Drainage	
	Disposal	
OTHER SERVICES		
General	Local authority	
	Police	
	Hospital	
Suppliers		
	Builders merchants	
	Ready-mix concrete	
	Quarries	
	Others specified	
Tips		
	Distance	
	Charges	
	Exclusions	
TEMPORARY WORKS		
Offices/storage	Position	
	Services	
	Movement	
Access roads/hard standing		
	Existing	
	New	
Signs		
	Directional	
	Advertising	
Existing buildings		
	Protection	
	Demolition	

QUALITY ASSURANCE

HEADINGS	REMARKS
GENERAL DETAILS	
Mileage to office	
Projects in area	
Local s/contractors	
Local plant supply	
General weather	
Exposure index	
Rainfall pattern	
Temperatures	
Type of adjacent buildings	
Residential	
Industrial	
Commercial	
Others (specify)	
NAMES/VISITING PARTIES	
Estimator	
Site staff	
Planner	
Contracts manager	
Project co-ordinator	
Others (specify)	
SKETCHES/PHOTOS	
List	
ANY OTHER DETAILS	
PREPARED BY	
Name	Position
Signature	Date

QA MANUAL	Initial Review Meeting	SECTION		Supporting Documentation	
		STAGE		Tender	
		REF	QAM 5.1	QA/E	
		REV		DATE	
		SHEET	1	of	1

AGENDA

Project name		Tender Ref	
Date of initial review meeting		Tender date	
Present			

ESTABLISHMENT OF TIMETABLE

Latest date for	Date	Action	Latest date for	Date	Action
Despatch of enquiries			Finalise programme		
Receipt of quotations			Contract Review Meeting		
Visit to site			Category Review Meeting		
Finalise method statement			Prog. and Method Review Meeting		
Complete cost estimate			Tender Review Meeting		

QUALITY OF TENDER INFORMATION

Drawings	Comment	Action	**Bills of Quantities**	Comment	Action
Location			Preliminaries		
Layout			Preambles		
Construction:			Measured work		
Architectural			Sketches		
Structural			**Specification**		
M&E					
Others					

TENDER INCEPTION DISCUSSION

Schedule of trades to be sub-contracted

Major materials and plant items

Site set up/staffing

ADDITIONAL ITEMS

Prepared by	Position	Date

CHECKING PROCEDURES - NEW FIRMS : Firms under Category C
(or A where applicable)

Project name	Ref
Company name	Address
Product or service to be supplied	
	Date
Checked by	

Note: If successful the company's name must be added to the Central Data Bank and a copy of this form filed in their records as well as the Tender/Contract Records File.

SUPPLIER	SUBCONTRACTOR
Points to establish	
(1) Do the Company's products comply with a recognised / specified standard? I.e.	(1) Does the Company's work comply with a recognised / specified / approved standard? I.e.
Tick	
a. has a BS Kitemark/CE Mark	Tick
b. has a BBA Certificate	a. a registered QA System ISO 9000
c. has a Test Certificate from other body	b. other QA System - 1st or 2nd party certified
d. will provide own written warranty	c. suitable confirmed 3rd party references
e. inspection and written assurance	d. successful inspection of previous work
f. other - specify	e. other - specify
(2) Do they manufacture to a consistent quality?	*Note- if the S/c is also supplying materials he will be required to check and verify conformance of*
Tick	*his materials in accordance with the criteria*
a. a registered QA System ISO 9000	*listed in the Materials column as a condition of*
b. other QA System - 1st or 2nd party	*their order*
c. inspection visits of previous projects	
d. other - specify	

Reason for conformance should be entered in the 'checked' column of QA / C

Example	
Supplier	**Subcontractor**
Conformance may therefore be entered	Conformance may therefore be entered
in the column as:	in the column as:
1(a) / 2(b)	**1(a)**

NOTE - it is important that copies of documents establishing conformance above must be obtained by the Area Office Contracts Dept. whilst placing the order.

QA MANUAL	Contract Review	SECTION	Supporting Documentation	
		STAGE	Tender	
		REF	QAM 5.1	QA/G
		REV		DATE
		SHEET	1	of	1

AGENDA	
Project name	Ref
1. Details of client or holding company	
2. Full particulars of consultants to be used including duties and responsibilities	
3. Particulars of site supervision to be provided by client/consultants	
4. Has tender been subject to a previous invitation	
5. Number of tenders to be invited	
6. Date for commmencement (actual)	
7. Details of any phasing	
8. An outline of method of construction	
9. Access problems	
10. Details of Nominated S/c or Suppliers	
11. Amendments to Standard Form of Contract	
12. Bonding requirements	
13. Insurance requirements	
14. Stage of design (i) today (ii) at commencement	
15. Any particular requirements/problems	
16. Unclear specification items	
17. Unclear drawing details	
18. Unclear Bill of Quantities items	

QA MANUAL	Tender Review Record	SECTION	Supporting Documentation	
		STAGE	Tender	
		REF	QAM 5.1	QA/H
		REV		DATE
		SHEET	1	of	1

PARTICULARS

Project name	Description
Date received	Date due Time
Tender to be delivered to	

ATTENDANCE

INFORMATION

Enquiry Report	
Form B1	
Form B2	

DISCUSSION

Form B1	
Form B2	
Overheads and profit	

DECISIONS

Record of decisions taken	Action by

QA MANUAL	Appointment Procedures	SECTION	Supporting Documentation	
		STAGE	Tender/Mobilisation/Construction	
		REF	QAM 5.1	QA/K
		REV	DATE	
		SHEET	1	of	1

Project name	Ref

GENERAL NOTE

If Consultants are to be chosen via these procedures the information necessary for Section B will be derived either

(1) from records of a previous interview and telephone conversation or

(2) interview specifically for this project

SECTION A

Scope of work	Company A	Company B	Company C

SECTION B

	Company A	Company B	Company C
Experience of project type			
Experienced with Client			
Requirements			
a. timetable			
b. fee			
c. quality *			
d. insurance			
e. contract			

SECTION C

Details of agreement			

NOTES

* Quality requirements either (1) QA System to ISO 9000 or (2) own QA System or (3) inspection visit

QA MANUAL	Handover Meeting Record	SECTION	Supporting Documentation	
		STAGE	Mobilisation	
		REF	QAM 5.1	QA/L
		REV		DATE
		SHEET	1	of	1

PARTICULARS

Project name	Site address
Meeting date	Site Telephone No.

ATTENDANCE

Information handed over	No. copies	Action		No. copies	Action
Enquiry Report			Bills of Quantities		
Site Visit Report			Tender Drawings and Spec		
Form B1			Net and Gross Printouts		
Form B2			Prelims Build-up		
Tender Review Record			Rates Build-up		
Enquiry Record			S/c and Supplier Quotes		
			Resourced Tender Prog.		

POINTS OF DISCUSSION / Notes

Contract conditions	
The site	
Method and programme	
Tender build-up	
Client/information appraisal	
Other	

Prepared by:	Position	Date

Project name		Ref	
Date of meeting		Site commencement date	

ATTENDANCE

MOBILISATION TIMETABLE

Item	Latest date	Action by	Item	Latest date	Action by
Check/arrange services			Finalise main programme		
Arrange site photos/survey			Prepare detail sub-progs.		
Open local accounts			Prepare initial histograms		
Arrange site offices, store etc.			Analyse temp. works required		
Review method of work			Prepare Quality Plan		
Prepare cost monitoring info.			Pre-commence client meeting		
Establish safety policy			Pre-commence co-ord. meeting		
Set up doc. control					

QUALITY OF CONSTRUCTION INFORMATION

Item	Comment	Date for receipt	Action by
Drawings			
Location			
Layout and elevation			
Construction			
Architectural			
Structural			
M&E			
Others			

SUPPLIERS/SUBCONTRACTORS

Schedule of trades to be subcontracted (Initial Review)

Trade	Action by	Trade	Action by	Trade	Action by

Schedule of materials and plant to be ordered (Initial Review)

Item	Action by	Item	Action by	Item	Action by

Site set up/staffing

Item	Action by	Item	Action by	Item	Action by

Other

Prepared by:	Position	Date

QA MANUAL	Contract Review (Stage 2)	SECTION	Supporting Documentation	
		STAGE	Mobilisation	
		REF	QAM 5.1	QA/N
		REV		DATE
		SHEET	1	of	1

AGENDA FOR CONTRACT REVIEW MEETING WITH CLIENT
(STAGE 2 PRE-COMMENCEMENT) * Refer to QA/G

Project name   Ref

Date of meeting   Venue

ATTENDANCE

Contractor	Client	Position	Tel No.

INFORMATION FROM CLIENT	INFORMATION FROM CONTRACTOR
Date for possession	Project team
Schedule of tests/samples	Construction programme
Bond	Quality Plan (initial period)
Insurances	Office accommodation
Particular client procedures	Date for commencement
Nominations	Particular company procedures
Provisional sums	
Contract signature/completion	

DRAWINGS/SPECIFICATION AVAILABLE/IN PREPARATION

Item	Ref No.	Rec. date	Anticipated date

FURTHER INFORMATION

Unclear specification items	
Unclear drawing details	
Unclear BQ details	
Date of next meeting	Venue

Prepared by: Position   Date

QA MANUAL	Pre-commencement Co-ordination Meeting	SECTION	Supporting Documentation	
		STAGE	Mobilisation	
		REF	QAM 5.1	QA/P
		REV		DATE
		SHEET	1	of 2

Project name		Ref	
Date of meeting		Venue	

ATTENDANCE

PROGRESS REVIEW MOBILISATION TIMETABLE

Item	Complete	Action by	Item	Complete	Action by
Check/arrange services			Finalise main programme		
Arrange site photos/survey			Prep. detailed sub-progs.		
Open local accounts			Prepare initial histograms		
Arrange site offices, store etc.			Analyse temp. works required		
Review method of work			Prepare Quality Plan		
Prepare cost monitoring inf.					
Establish safety policy					
Set up doc. control					

QUALITY OF CONSTRUCTION INFORMATION (UPDATE AND REVIEW)

Drawings received	Comments	Action by

Drawings awaited	Due by date	Action by

RESULTS OF EXAMINATION OF DRAWINGS/SPECIFICATION

Items requiring architect's approval/action

Item	Ref	Item	Ref	Item	Ref

Items which are unusual/problems

Item	Ref	Item	Ref	Item	Ref

SUPPLIERS/SUBCONTRACTORS

Trades/item - ordered	Trades/item - enquiries sent	No. sent	No. retnd	Action by

SITE/PROJECT MANAGEMENT

Discuss/complete staff organisation chart | Action by

Staff job descriptions issued (*see Standard Descriptions QA Manual **)*

Name	Position	Name	Position	Name	Position

Prepared by: | Position | Date

QA MANUAL	Site (co-ordination) Meeting Timetable	SECTION		Supporting Documentation		
		STAGE		Construction		
		REF	QAM 5.1	QA/Q		
		REV		DATE		
		SHEET	1	of		1

Project name	Ref
Name of subcontractor/consultant	
Address	Telephone No.
Representative to contact	Extension No.
Venue for meetings	
Time of meetings	
Contractor's representative to contact	Telephone No.
Directions to venue	

TIMETABLE

Date for possession	S/c commencement date
Date for completion	S/c completion date
E.O.T. granted to main contractor	EOT granted to S/c
Details of phasing	Details of phasing

Meeting No.	Date	S/c required to attend	Meeting No.	Date	S/c required to attend
1			22		
2			23		
3			24		
4			25		
5			26		
6			27		
7			28		
8			29		
9			30		
10			31		
11			32		
12			33		
13			34		
14			35		
15			36		
16			37		
17			38		
18			39		
19			40		
20			41		
21			42		

Note: Please confirm your ability to attend

* Revisions to the timetable to be issued if necessary on this form

QA MANUAL	Site Construction Co-ordination Meeting	SECTION	Supporting Documentation	
		STAGE	Construction	
		REF	QAM 5.1	QA/R
		REV	DATE	
		SHEET	1	of

AGENDA

Project name   Ref

Meeting No.   Venue   Date

Present

Agree minutes of previous meeting	Yes	No

Errors or omissions

Record of inclement weather in period   to date

Detailed progress report related to programme and agreed with client's representative (see Sheet 3)

Check action required from previous minutes

List items dealt with	Action
List items outstanding One month	
Longer	

Instructions issued in the period

Drawings/information issued in the period

C.O.W. Directions issued in the period

REQUESTS FOR INFORMATION IN PERIOD

Ref No.	Date sent	Date replied	Date part replied	Info. outstanding	No reply

REQUESTS FOR INSPECTION IN PERIOD

Ref No.	Date	Result	Date	Action

DELAY NOTIFICATIONS TO DATE

Contractor

Ref No.	Date	Reply received	Action required

Subcontractor

Ref No.	Date	Reply received	Action required

Any other business	Action

Date for next meeting if not as scheduled	Venue if not as scheduled

Prepared by:	Position	Date

QA MANUAL	Site Construction Co-ordination Meeting (contd)	SECTION	Supporting Documentation	
		STAGE	Construction	
		REF	QAM 5.1	QA/R
		REV		DATE
		SHEET		of

Progress listing % complete against programme operations

Operation	Programme %	Actual %

Prepared by	Position	Date

QA MANUAL	Site Subcontractors Co-ordination Meeting	SECTION	Supporting Documentation	
		STAGE	Construction	
		REF	QAM 5.1	QA/R(s)
		REV		DATE
		SHEET	1	of

AGENDA

Project name	Ref	
Meeting No.	Venue	Date

Present

Agree minutes of previous meeting	Yes	No

Errors or omissions

Record of inclement weather in period	to date

Subcontractors' progress reports related to programme and agreed with contractor's representative (see Sheet 3)

Check action required from previous minutes

List items dealt with	Action

List items outstanding	One month	

Longer	

Subcontract instructions issued in period to each S/c (S/c listed in alphabetical order) - enter latest reference number

Name	Ref no.	Name	Ref no.

Drawings issued in period (S/c listed in alphabetical order)

S/c Quality Plans

Name	Up to date?	Action

Test/Inspection certificates supplied in period (S/c listed in alphabetical order)

Name	No.	Name	No.

Work stoppage certificates issued - this period

Work stoppage certificates issued - previous period

Work stoppage certificates released

Any other business	Action

Date for next meeting if not as scheduled	Venue if not as scheduled

Prepared by:	Position	Date

QA MANUAL	Site Subcontractors Co-ordination Meeting (contd)	SECTION		Supporting Documentation	
		STAGE		Construction	
		REF	QAM 5.1	QA/R(s)	
		REV		DATE	
		SHEET		of	

Progress listing % complete against programme operations

Operation	Programme %	Actual %

Prepared by:	Position	Date

Project name		Month	Year	
Ref		Date prepared		

PROGRESS

Programme ref		Revision		Date

Progress listing % complete against programme operations

Operation No.	Programme %	Actual %	Operation No.	Programme %	Actual %

Note numbers here of operations which were completed during previous period

Contract completion date	EOT claimed	*Days
Revised completion date	EOT granted	*Days
Estimated completion date	EOT not claimed	*Days
* Working days	EOT rejected	*Days

RESOURCES

Labour	Approx hours/value resourced	Approx hours/value expended
Own labour		
S/c labour		
Own staff		
Agency staff		

QA MANUAL	Monthly Contract Report (contd)	SECTION	Supporting Documentation		
		STAGE	Construction		
		REF	QAM 5.1	QA/T	
		REV		DATE	
		SHEET	2	of	4

RESOURCES (CONTD)

Major plant	Approx hours resourced to date	Approx hours expended to date

Mark own (o) or hired (h)

Major subcontractors	Programme		Comments	No. QA/W certs issued in period
	OK	(+) / (-)		

RESOURCES (CONTD)

Major materials in period	Quantity in valuation	Quantity delivered	Quantity in stock	% waste

INFORMATION

	No. issued		Answered		Unanswered	
	To date	Period	To date	Period	To date	Period
Information requests (QA/U)						
Information appraisal (QA/V)						
Inspection requests (QA/W)						

	No. issued	
	To date	Period
Confirmation of verbal instructions (QA/X)		
Architects/SO instructions		
C.O.W directions		
Subcontractor/supplier instructions		

Drawings

	Latest ref No.	No. in period
Architects/SO		
Subcontractors		
Suppliers		
M & E		
Others		

INFORMATION (CONTD)

	Latest Ref No.	No. in period
Letters from architect		
Letters to architect		
Letters from S/c/suppliers		
Letters to S/C/suppliers		
Others (list)		

Resource problems - (note particular problems including information)

Enclose copies of the following

- Final account progress report
- Loss damage or injury report
- Diary pages for period
- Test/inspection certificates
- Minutes of latest site co-ordination meetings
- Estimator feedback Form QA/Z
- Others (list)

SITE MANAGER'S COMMENTS

Prepared by   Position   Date

QA MANUAL	Information Request	SECTION	Supporting Documentation	
		STAGE	Construction	
		REF	QAM 5.1	QA/U
		REV		DATE
		SHEET	1	of	1

Project name   Ref

Date

Description

Date required by   Signed

Reply

Date   Signed

QA MANUAL	Information Receipt/ Appraisal	SECTION	Supporting Documentation		
		STAGE	Construction		
		REF	QAM 5.1	QA/V	
		REV		DATE	
		SHEET	1	of	1

Project name	Ref
Document received	
Document ref. (from DC /1)	
Date	
Checked by	

OK to proceed	Query or extra information

Information required

Date required by

Signed   Date

QA MANUAL	Work Inspection Certificate	SECTION		Supporting Documentation	
		STAGE		Construction	
		REF	QAM 5.1		QA/W
		REV		DATE	
		SHEET		of	

Subcontractor or own work

Project name		Ref			
Date					
Subcontractor name		Person responsible			
Operation number		Description			

Hold points	Description	Proceed date/ Signature	Hold date/ Signature	Released date/ Signature

Waivers

Hold points	Description of waiver	Date

QUALITY ASSURANCE

QA MANUAL	Confirmation of Receipt of Verbal Instructions	SECTION	Supporting Documentation		
		STAGE	Construction		
		REF	QAM 5.1	QA/X	
		REV		DATE	
		SHEET	1	of	1

Date

Verbal instruction from _____ Date _____

Description of instruction

Forward to architect within seven days of verbal instructions

Prepared by: _____ Position _____ Date _____

NOTE - Reference any follow up correspondence from architect here

QA MANUAL	Material/Plant Confirmation Receipt	SECTION		Supporting Documentation	
		STAGE		Construction	
		REF	QAM 5.1	QA/Y	
		REV		DATE	
		SHEET		of	

Project name   Ref

Date

Brief Description	Delivery note number	Date	Inspected by *	Comments

* Authorised persons nominated by the site manager only

Project name		Ref	
Report No.	Prepared by	Date	

Subcontractor	Performance rating	Comments Goods reject notes, inspection records, work reject notes, rectification process, re-inspection

Performance rating - * Very Poor ** Poor *** Satisfactory **** Good

QA MANUAL	Estimator Feedback	SECTION	Supporting Documentation	
		STAGE	Construction	
		REF	QAM 5.1	QA/Z
		REV		
			DATE	
		SHEET	2	of	3

Supplier	Performance rating	Comments Goods reject notes, inspection records, work reject notes, rectification process, re-inspection
Consultants	**Performance rating**	**Comments**

Performance rating - * Very Poor ** Poor *** Satisfactory **** Good

Rate build-up variances

BQ item	Rate (nett)	Cost	Any special circumstances	Other comments

This page left intentionally blank

12.0   Generally

Having described the process of implementing a QA system and examined its constituent parts you will have observed that it is a complex and lengthy process. This is not the end of the development of the system.

It is after all the time and effort expended in producing the initial system that a dangerous point is reached. The danger is that once the system is **complete** it will become a neatly packaged inviolate document. This is the very thing that must be avoided and positive action must be taken to prevent it. A well directed auditing programme should be focused upon making the procedures more effective in terms of both quality and the company's aims. However it is executive management's role within the system to ensure its maintenance and a sure sign for an external auditor that a system is not working is the absence of change.

QA is about getting it right first time but this is an aim that will forever remain on the horizon; the important thing is that the journey is in the right direction. Having prepared the manual a culture must be developed by the company within the individuals who use the system to question and improve the procedures used. Both management and individuals must aim for improvement and appreciate that evolutionary change is essential to take the path that leads in the direction of getting it right first time.

For the purposes of obtaining ISO 9000 certification it is wise not to include more than is necessary within the audited procedures. From the alternative viewpoint of improving the organisation's efficiency it makes sense that all areas of the company should eventually be encompassed within a broader system. The discipline inherent in a QA system can improve the way all activities are undertaken. The requirements of the British Standard draw in those activities that have a primary relationship with the finished product. Activities such as marketing and accountancy may have only a secondary relationship but their efficiency is important to the organisation as a whole.

To conclude therefore it may be helpful to abstract the key points brought out in the previous chapters.

12.1   Summary of 24 key points

Point 1
QA is not an optional extra but a philosophy that management must be committed to.

◆ Chapter 1

Point 2

A QA system to have any credibility must comply with ISO 9000.

◆ Chapters 1 and 3

Point 3

The production of a QA system encounters two major problems:

- the volume of work it requires
- blending a knowledge of QA and the organisation.

◆ Chapter 1

Point 4

The system should reflect the company's objectives and not simply address the requirements of ISO 9000.

◆ Chapter 9

Point 5

Involve employees early in order to:

- avoid creating unnecessary concern or antipathy
- instigate a sense of property.

◆ Chapter 3

Point 6

Use graphics as an aid to conveying information.

◆ Chapter 3

Point 7

Note the aspects of a good system illustrated by Rudyard Kipling's six servants.

◆ Chapters 3 and 4

Point 8

The QA manual should be established as the company's main reference document.

◆ Chapter 3

Point 9

Construction based QA systems operate at two levels:

- company wide
- project specific.

◆ Chapters 2, 4 and 8

Point 10

The Documented Quality System describes the way an organisation conducts its business.

◆ Chapter 2

Point 11

The manual, or relevant section thereof, must be available to all with roles or responsibilities within it.

◆ Chapter 2

Point 12

Distribution of the manual and amendments must be controlled and recorded.

◆ Chapters 2 and 4

Point 13

Company organisation charts are best developed in two tiers:

- central administrative organisation
- project based organisation.

◆ Chapter 4

Point 14

The manual is simplified if developed relative to the different groups involved:

- tender
- mobilisation
- construction stages.

◆ Chapter 4

Point 15

Describing what is done is extremely complex so layer the information e.g. flow charts, procedures and forms.

◆ Chapters 4-7

Point 16

Meetings should be kept brief but are a useful aid to communication.

◆ Chapter 4

Point 17

Time, resources, information and control pose problems at all stages which must be addressed and controlled.

◆ Chapters 4-7

Point 18

The Quality Plan is completed pre-construction but tiering again helps giving:

- overall plan
- necessary detail.

◆ Chapter 8

Point 19

Internal audits should aim to identify ineffective procedures, with the purpose of improving quality in terms of the company's objectives.

◆ Chapter 9

Point 20

Internal audits check the system not the work; responsibility remains with management at all levels.

◆ Chapter 9

Point 21

Employees are an important bought-in resource controlled via recruitment and training.

◆ Chapter 9

Point 22

Communication equates with good management and should be incorporated wherever possible within the QA system.

◆ Chapter 3

Point 23

Audit programmes should aim to cover the whole company once a year.

◆ Chapter 9

Point 24

The basic role of the internal auditor is to find out what is actually happening within the company.

◆ Chapter 9

Finally remember your QA system is a live thing that must encourage a two-way flow of information as it must change to improve. To quote from Sir Winston Churchill:

There is nothing wrong with change if it is in the right direction. To improve is to change, so to be perfect is to have changed often.